Sustainable Production, Life Cycle Engineering and Management

Series editors

Modern production enables a high standard of living worldwide through products and services. Global responsibility requires a comprehensive integration of sustainable development fostered by new paradigms, innovative technologies, methods and tools as well as business models. Minimizing material and energy usage, adapting material and energy flows to better fit natural process capacities, and changing consumption behaviour are important aspects of future production. A life cycle perspective and an integrated economic, ecological and social evaluation are essential requirements in management and engineering. This series will focus on the issues and latest developments towards sustainability in production based on life cycle thinking.

More information about this series at http://www.springer.com/series/10615

Supachai Vongbunyong · Wei Hua Chen

Disassembly Automation

Automated Systems with Cognitive Abilities

Springer

Supachai Vongbunyong
School of Mechanical and Manufacturing Engineering, Sustainable Manufacturing and Life Cycle Engineering Research Group
University of New South Wales
Sydney
Australia

Wei Hua Chen
School of Mechanical and Manufacturing Engineering, Sustainable Manufacturing and Life Cycle Engineering Research Group
University of New South Wales
Sydney
Australia

ISSN 2194-0541 ISSN 2194-055X (electronic)
Sustainable Production, Life Cycle Engineering and Management
ISBN 978-3-319-15182-3 ISBN 978-3-319-15183-0 (eBook)
DOI 10.1007/978-3-319-15183-0

Library of Congress Control Number: 2015932072

Springer Cham Heidelberg New York Dordrecht London

Printed on acid-free paper

Springer International Publishing AG Switzerland is part of Springer Science+Business Media (www.springer.com)

Dedicated to Our Families
Arpa and Sumeth Vongbunyong,
Prapassri Leekphai

—Supachai Vongbunyong

Preface

As the world's population exponentially grows, consumption rates and the demand for new products also increase dramatically. As a consequence, a great number of end-of-life (EOL) products are continuously being disposed of, leading to a number of environmental problems. Responsible EOL treatment—which may include reusing, recycling or remanufacturing products or parts—is desirable in dealing with these disposed products. These processes can be beneficial both environmentally and economically. Waste is minimised, while valuable components and materials are recovered.

The *disassembly of products* is one of the primary steps of EOL treatment processes, and involves the extraction and segregation of the desired components, parts or materials from the product. Disassembly does not only input towards EOL treatment, but also allows the repair and maintenance of products. However, most of this process is economically infeasible due to time consumption, process difficulty and expensive labour costs. Consequently, the option of disassembly is often ignored in industry.

Replacement of human labour by automation has been successful in increasing the cost-effectiveness of many industries, especially manufacturing and production processes. Therefore, the implementation of an automated system in the disassembly process is considered as one possible solution. However, the disassembly process involves a number of challenging problems and cannot be considered as the reversal of the assembly process. A number of difficulties arise due to three main aspects: the physical uncertainties associated with the end-of-life product condition, the large variety within the one product category, and complexities in process planning and operation. Therefore, disassembly automation needs to be designed to be flexible and is robust enough to overcome these issues.

This book provides an overview of the design of disassembly automation, along with a case study example of the development of a new system based on the research, "Cognitive robotics in the disassembly of products", conducted at *the University of New South Wales, Australia*. The general concept of product disassembly is introduced and a review of the existing disassembly automation systems is presented. After that, the book provides an overview of the general system set-up, followed by detail into each primary operating module of the automated system. This book is organised as follows.

Chapter 1 describes the importance of product disassembly as a key step in the end-of-life treatment process. This chapter also presents an overview of the current research direction in the field of disassembly.

Chapter 2 provides an overview and literature review of the disassembly process. The literature shows that a number of techniques have already been developed at the planning and operational levels, typically for optimising the disassembly process for economic feasibility. These techniques can be implemented in both manual and autonomous disassembly.

Chapter 3 considers the disassembly system as the integration of a number of operating modules working together to achieve the goal. An overview of this configuration is described. Existing research regarding the development of a (semi-) autonomous disassembly system and disassembly tools is reviewed. In addition, the set-up of the workstation and system framework used in this research is explained.

Chapter 4 provides an overview of perception in the disassembly system. Detection techniques, in regard to hardware and software used in existing research, are reviewed. This chapter also describes the implementation of the vision system in this research, including the detection of components based on common features and coordinate mapping using the depth camera.

Chapter 5 explains the principle of cognitive robotics. The cognitive robotics agent is an intelligent planner that controls the behaviour of the system in order to overcome the variations and uncertainties in the disassembly process. The behaviour is influenced by four cognitive functions, namely reasoning, execution monitoring, learning and revision.

Chapter 6 describes the integration of the aforementioned operating modules into a complete disassembly system. The software system applies the vision system, operation plans and the principle of cognitive robotics to a disassembly cell specifically designed for disassembling LCD screens. The detailed configuration of the system and additional information specific to the case-study product are also explained.

Chapter 7 presents the conclusions developed as a result of this research in the development of a disassembly automation system. Technical perspectives of the system, its economic feasibility and the future work are also presented.

Acknowledgments

First, we would like to thank our supervisors Profs. Sami Kara and Maurice Pagnucco for the great opportunity given to us to work on this exciting research topic. They have always given the best support in terms of research direction, theory and technical help, which have been crucial in producing this work.

Next, we would like to thank the School of Mechanical and Manufacturing Engineering for their provision of research funding and facilities. In addition, we would like to thank the workshop and technical staff members, Martyn, Seetha, Russell, Alfred, Ian, Andy, Subash, Radha and Steve, for their great technical support and the manufacture of the hardware parts. We would like to thank Drs. Erik van Voorthuysen and David Rajaratnam (CSE) for their valuable suggestions and comments in the early stage of the disassembly cell set-up in regard to robotics and programming. In addition, we would like to thank TAD NSW Disability Services for supplying and donating LCD screens for testing.

We would also like to thank the members of the Sustainable Manufacturing and Life Cycle Engineering Research Group (SMLCE@UNSW), in particular Dr. Suphunnika Ibbotson, Dr. Wen Li, Dr. Seung Jin Kim, Dr. Bernard Kornfeld, Dr. Kanda Boonsothonsatit, Dr. Rachata Khumboon, Pouya Ghadimi Karahrodi, SeyedHamed MoosaviRad, Smaeil Mousavi, Wei Lau, Samira Alvandi and Scott Ibbotson, for the sharing of ideas, valuable comments, their warm welcomes and all the other assistance that they have provided. Moreover, we would like to thank our German colleagues, in particular Prof. Dr.-Ing. Christoph Herrmann, Dr. Tobius Luger, Gerrit Bogdanski and other researchers from JGARG for their insights into the disassembly of LCD screens, LCA and manufacturing.

I, Wei Hua, would additionally like to thank the crew at UNSW Mechatronics, in particular Dr. Mark Whitty, Dr. Jose Guivant, Dr. Ngai Kwok and Michael Woods for the technical and personal support they have provided, and all I have learnt in our various conversations. Thanks also to the MoFA group at the Institute of Machine Tools and Production Technology (IWF), TU Braunschweig, where I stayed while writing and editing a part of this book, for their welcome reception and support, as well as their ready provision of resources. Thank you to my family, my friends and God, for bringing me to the point where I am now.

I, Supachai, would also like to thank most importantly my family, particularly my beloved father, mother and wife for the lifelong support given me at all times.

Supachai Vongbunyong
Wei Hua Chen

Contents

Abbreviations

AI	Artificial Intelligent
CCFL	Cold-Cathode Fluorescent Lamp
CRA	Cognitive Robotic Agent
CRM	Cognitive Robotic Module
DOF	Degree of Freedom
DOM	Disassembly Operation Module
DPP	Disassembly Process Plan
DSP	Disassembly Sequence Plan
EOL	End-of-Life
GUI	Graphic User Interface
LCA	Life Cycle Assessment
LCE	Life Cycle Engineering
MAS	Multi-Agent System
MBR	Minimum Bounding Rectangle
PCB	Printed Circuit Board
ROI	Region of Interest
VOI	Volume of Interest
VSM	Vision System Module
WEEE	Waste Electrical and Electronic Equipment

Chapter 1
Introduction

Abstract End-of-life (EOL) treatment is a main stage in the life cycle of a product, and often aims to recover the remaining embodied value of the disposed products. Disassembly is a key activity in efficient EOL treatment. This chapter presents the general idea of the disassembly of products in view of EOL treatment. An introduction of the current research interest in this field, especially disassembly automation, is provided.

1.1 End-of-Life Product Treatment

Life Cycle Engineering (LCE) is an engineering discipline that focuses on a systematic approach to designing a product, considering its entire life cycle and incorporating the environmental aspects along with the economic, technical and social aspects during the product development [1]. The life cycle of products consists of four stages, namely the (a) *material stage*, (b) *manufacturing process*, (c) *usage*, and (d) *end-of-life (*EOL*)*, as shown in Fig. 1.1. The analysis of the entire life cycle is known as a "*cradle-to-grave*" analysis. In the beginning stages, the raw materials are produced by the material suppliers and then conveyed to the manufacturing process for production. After manufacturing, the products are distributed to the market and used by the consumer during the usage stage. Finally, the products are disposed of in the end-of-life stage.

EOL products disposed of in an inappropriate manner can result in environmental problems. EOL products that contain hazardous substances, such as electronic waste, require special treatment options. One can also attempt to recover the value remaining in the products; this recovery can take many forms, e.g. energy, materials, components, even entire products. These outcomes can be fed back into other stages of the product life cycle. This provides benefits not only in the environmental but also the economic aspect.

This book focuses on the EOL stage. Products arrive at the EOL stage in differing conditions according to the conditions experienced during the usage stage.

S. Vongbunyong and W.H. Chen, *Disassembly Automation*, Sustainable Production, Life Cycle Engineering and Management, DOI 10.1007/978-3-319-15183-0_1

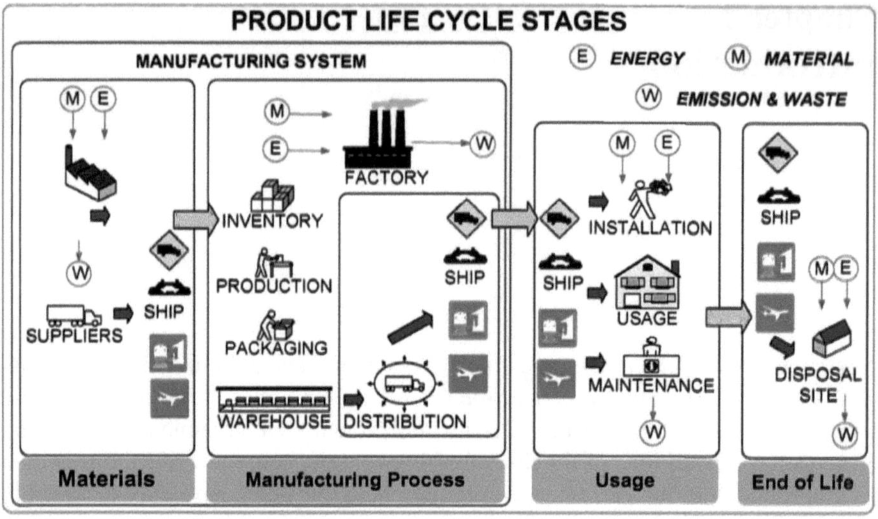

Fig. 1.1 Product life cycle stages [2]

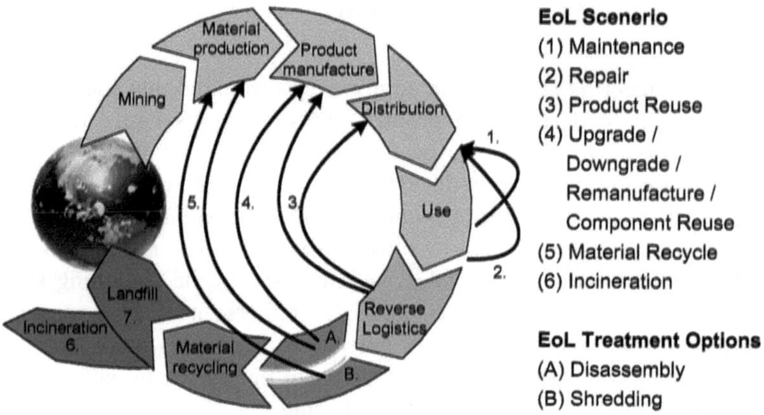

Fig. 1.2 Scenario of end-of-life products [3]

After the usage stage, the disposed EOL products are collected by the reverse logistics for appropriate treatment. Value recovery from EOL products in EOL treatment can be achieved by three major activities, namely (a) *reusing*, (b) *remanufacturing*, and (c) *recycling*. The most appropriate process can be decided by considering the environmental and economic aspects, which can be analysed based on the Life Cycle Assessment (LCA) method.

From Fig. 1.2, EOL products that are in good condition can be immediately *reused* by distributing them back to the customers. These may also be sent back to the manufacturers for refurbishment. Non-functional EOL products can be considered for *remanufacturing* or *recycling*. Before these treatment options can be carried out, the products[1] must be disintegrated into subassemblies,[2] components,[3] or materials according to the requirements of each treatment option. Options for disintegration include *disassembly* and *shredding*.

Disassembly systematically separates the product into its components and/or sub-assemblies [4]. The outcomes of disassembly can supply a wide range of treatment options, according to the desired purpose and downstream condition. Apart from EOL treatment, disassembly also serves the purpose of repair-maintenance if appropriate disassembly techniques are applied. A disassembly process can preserve the value of detached parts, providing neat and high quality outcomes suitable for refurbishment or reuse. However, disassembly processes are typically expensive due to the difficulty of the task and the demand for manual labour.

Shredding is a destructive process that breaks the products into small pieces or particles, eventually to be supplied to recycling processes. The outcome of shredding is a low-quality blend of material which requires a subsequent sorting process to separate the valuable material from the scraps. The material blend is sorted via their physical properties using a number of techniques e.g. magnetic, electrostatic, and eddy-current separation. Shredding is commonly implemented in industry due to the low operating cost. However, a major disadvantage is the loss of value of the parts and components. In addition, the shredding of components containing hazardous substances may also be problematic due to the contamination of the workplace and other materials in the process [5].

From Fig. 1.2, for *remanufacturing*, products are disintegrated into sub-assemblies, components, or parts, with minimal damage, in order to retain their functionality. The embodied value from manufacturing and the materials are hence recovered. On the contrary, for *recycling*, only the value of the materials is recovered. Disassembly can also be advantageous in obtaining a purer material fraction for recycling, with less sorting processes required.

In summary, the outcome of the disassembly process can be supplied to various EOL treatment processes, as summarised in Table 1.1. In comparison to the shredding process, disassembly tends to produce an outcome that is in better condition and better retains the embodied value of the components. The high operating cost often exceeds the value recovered from the EOL products; hence, it becomes economically infeasible and is usually avoided in industry practice. For making disassembly economically feasible, much research has been conducted to develop the strategies and tools explained in the following sections.

---

[1] *Product* performs a particular functionality and consists of a number of discrete components and/or subassemblies connected together.

[2] *Subassembly* is a group of components connected together. It can be considered a *module* if the subassembly performs a particular functionality.

[3] *Component* refers to a discrete part that cannot be further disassembled. (See detail in Sect. 2.1.2).

Table 1.1 Destination of output of a disassembly facility (Adapted from [5])

Process outcome	Outcome of disintegration			Destination to EOL treatment process				
	No process	Disassembly	Shredding	Refurbish	Remanufacture	Reuse	Recycle	Incinerate/Landfill
Unprocessed products	•			•			•	•
Modules or sub-assembly		•		•	•		•	•
Components		•				•	•	•
Damaged components		•					•	•
Waste			•					•

1.2 Disassembly of Products

The EOL treatment process has become more of concern due to the increasing number of products disposed of globally. In the field of LCE, much research focuses on products which have a high material-return rate, short life-cycle, and high volume of waste, e.g. Waste Electrical and Electronic Equipment (WEEE) [6] and EOL vehicles. In 2005, the amount of WEEE in EU-27 was around 8.3–9.1 million tons per year, consisting 40 % of large household appliances and 25 % of medium-sized household appliances. This figure is expected to grow 2.5–2.7 % each year [7]. Moreover, the amount of WEEE is 8 % of the whole municipal solid waste (MSW) worldwide [8]. In 2006, around 230 million cars were in use in Europe (EU-15) with 10.5 million tons of vehicle waste being disposed of every year. According to the European Directive 2000/53/EC, at least 80 % by mass of the EOL vehicles must be reused and recovered; 85 % of this must be recycled [9].

A number of investigations have been conducted in regard to the environmental and economic aspects of EOL disassembly processes [10–12]. The disassembly approach has also been compared with conventional waste treatment options, i.e. disposal and landfill [13]. These investigations concluded that disassembly greatly benefits the environment but it is not an economical process due the excess *direct* and *indirect* costs. The direct costs are the costs involved in the disintegration of the product, relating to the labour costs, disassembly tools, disassembly system, etc. On the other hand, the indirect costs relate to the activities beyond the disintegration process, e.g. stocks and logistics. The disassembly process may become economically feasible if a more optimal disassembly strategy with respect to the costs and benefits can be implemented.

1.2.1 Research Directions

Ongoing research focuses on developing a strategy for economically feasible disassembly. Gupta and McLean [14] provide an overview of the research direction which can be categorised into four relevant areas: (a) *Design for Disassembly*, (b) *disassembly process planning*, (c) *design and implementation of disassembly systems*, and (d) *operations planning issues in the disassembly environment.*

Design for Disassembly (DfD) aims to resolve the difficulty in the disassembly process from the beginning, by designing the product to facilitate disassembly. This can bring a number of benefits, such as a reduction of the work needed in disassembly, quick and easy disassembly operations, simple product configuration, and easy handling. The guideline is presented in Boothroyd and Alting [15]. This area also includes *disassembly-embedded design* [16] and active disassembly [17].

Disassembly process planning deals with the development of rules, procedures, and software tools, used for formulating disassembly strategies and the disassembly system. Development of optimal disassembly sequence plans is one major theme in this area.

Design and implementation of disassembly systems deals with disassembly in the management level. This includes the methods used for establishing disassembly facilities and formulating economic and environmental evaluations of systems, as well as logistic networks for EOL treatment.

Operations planning issues in the disassembly environment deals with disassembly in the operational level. The operation level is affected by the logistics network and also concerns problems such as resource availability, collection and scheduling.

1.2.2 Automated Disassembly

In addition to these research areas, automated disassembly can also be a key in achieving future economic feasibility for disassembly. Currently, the disassembly process is generally conducted by human labour, which results in high operating costs, especially in developed countries. A number of attempts have been made to automate the process. However, variations and uncertainties in product and process have restricted the adoption of automated disassembly in industry to only a few strategic instances, where a manufacturer has selected to remanufacture its own products. Developing a system that is flexible and robust enough to overcome these difficulties, particularly as an entity separate from the product manufacturer, is a challenging problem and the theme of this book.

The automatic disassembly system can be considered a robotic system that employs an artificial intelligence (AI) agent controlling the mechanical operation units and sensors. The system achieves flexibility by perceiving relevant information during the process and adapting the operation accordingly. The development of these systems involves a number of engineering disciplines in addition to the field of disassembly:

- Product analysis;
- Disassembly process planning;
- Mechanical and control systems;
- Vision and sensor systems; and,
- Intelligent planners.

This book aims to presents the background knowledge and a case-study implementation according to these disciplines. The literature, regarding the existing research, theory and practice in the relevant areas, is also reviewed.

References

1. SMLCE (2013) What is Life Cycle Engineering, Sustainable Manufacturing & Life Cycle Engineering Research Group @ UNSW. http://www.lceresearch.unsw.edu.au
2. Kara S, Ibbotson S (2011) Embodied energy of manufacturing supply chains. CIRP J Manufact Sci Technol 4(3):317–323

3. Duflou JR, Seliger G, Kara S, Umeda Y, Ometto A, Willems B (2008) Efficiency and feasibility of product disassembly: a case-based study. CIRP Ann Manuf Technol 57(2):583–600
4. Kaebernick H, Ibbotson S, Kara S (2007) Cradle-to-cradle manufacturing. In: Transitions: pathways towards sustainable urban development in Australia, CSIRO Press, Australia, pp 521–536
5. Lambert AJD, Gupta M (2005) Disassembly modeling for assembly, maintenance, reuse, and recycling. CRC Press, Boca Raton
6. Parliament (2003) Directive 2002/96/EC of the European Parliament and of the council on waste electrical and electronic equipment (WEEE) of 27 Jan 2003
7. Jaco H, Federico M, Ruediger K, Claudia M, Clara D, Eniko A, Josef S, Ab S (2008) Final Report. Review of Directive 2002/96 on Waste Electrical and Electronic Equipment (WEEE). United Nations University
8. Babu BR, Parande AK, Basha CA (2007) Electrical and electronic waste: a global environmental problem. Waste Manage Res 25(4):307–318
9. Viganò F, Consonni S, Grosso M, Rigamonti L (2010) Material and energy recovery from Automotive Shredded Residues (ASR) via sequential gasification and combustion. Waste Manage 30(1):145–153
10. Chen KZ (2001) Development of integrated design for disassembly and recycling in concurrent engineering. Integr Manuf Syst 12(1):67–79
11. Gungor A, Gupta SM (1999) Issues in environmentally conscious manufacturing and product recovery: a survey. Comput Ind Eng 36:811–853
12. Li W, Zhang C, Wang HPB, Awoniyi SA (1995) Design for disassembly analysis for environmentally conscious design and manufacturing. In: ASME International Mechanical Engineering Congress and Exposition, pp 969–976
13. Ewers H-J, Schatz M, Fleischer G, Dose J (2001) Disassembly factories: economic and environmental options. In: IEEE international symposium on assembly and task planning, pp 447–452
14. Gupta M, McLean CR (1996) Disassembly of products. Paper presented at the 19th international conference on computers and industrial and engineering, Computer Industrial Engineering
15. Boothroyd G, Alting L (1992) Design for assembly and disassembly. CIRP Ann Manuf Technol 41(2):625–636
16. Masui K, Mizuhara K, Ishii K, Rose C (1999) Development of products embedded disassembly process based on end-of-life strategies. In: The EcoDesign'99: 1st international symposium on environmentally conscious design and inverse manufacturing, Tokyo, pp 570–575
17. Chiodo JD, Billett EH, Harrison DJ (1999) Active disassembly using Shape Memory Polymers for the mobile phone industry. In: IEEE international symposium on electronics and the environment, pp 151–156

Chapter 2
General Disassembly Process

Abstract The economic feasibility of the disassembly process is a main issue restricting its implementation in industry practice. Much research in the planning of disassembly processes and operations has been conducted in order to increase its economic feasibility. This chapter presents various aspects of the disassembly process including product representation, disassembly sequence planning (DSP), and dismantling techniques. This general knowledge is not limited to manual disassembly, but is also useful in automatic disassembly, which is presented in the following chapter.

2.1 Disassembly Process Planning (DPP)

The disassembly process is generally economically infeasible due to the difficulties in the process. Designing products according to Design for Disassembly (DfD) guidelines [1, 2] is expected to resolve this problem by making the disassembly process easier. However, few products nowadays are actually designed according to DfD. Therefore, the disassembly process remains difficult for the majority of products. Hence, this book focuses on the means of improving the economic feasibility of disassembly apart from DfD.

Duflou et al. [3] summarise the factors that influence profitability of the disassembly process. Two major factors which are further explained in this book are the (a) *completeness of disassembly* and (b) *degree of autonomy* of the process. The desired completeness or depth of disassembly is a question addressed in disassembly process planning and disassembly sequencing, and is further explained in Sect. 2.2. The degree of autonomy can vary from complete manual disassembly to semi-automatic disassembly and fully-automatic disassembly. Since automation is a major theme of this book, an overview is explained in Chap. 3, with other details presented throughout the rest of the book.

This chapter gives an overview and literature review regarding disassembly process planning.

S. Vongbunyong and W.H. Chen, *Disassembly Automation*, Sustainable Production, Life Cycle Engineering and Management, DOI 10.1007/978-3-319-15183-0_2

2.1.1 Difficulties in Disassembly

The disassembly process cannot be considered as the reverse of assembly. This is largely due to increased uncertainties: the disassembly process deals with unpredictable characteristics in both the quality and quantity of EOL products. This causes disassembly to be more difficult than assembly in the following aspects.

Uncertainties within models

Gungor and Gupta [4] summarise the physical uncertainties that can be found in EOL products manufactured under the same model. These uncertainties result from (a) *component defects*, (b) *upgrading or downgrading during usage* and (c) *damage during the disassembly operation.*

Defective main or connective components can result in difficulties during the removal operation which range from undesirable to dangerous. Examples include chemical leakage from batteries and broken fasteners that cannot be disestablished using common disassembly tools.

Upgrading and downgrading of the product during the usage stage can result in a change in product and component configuration. This situation is commonly found in devices containing exchangeable modules such as the personal computer (PC). Repairs and upgrades, e.g. involving the installation of random access memory (RAM) or graphics card, are common during the usage stage.

Damage in the disassembly process potentially occurs when the returned product is fragile. The disassembly process may need additional steps or a change in the disassembly sequence when certain parts are likely to break during the process.

Model-related variations

Products are manufactured into different models within the same product family. Different models contain variations in characteristics, including material, size and internal configuration. The same model may also be sold under different brands. Optimally, model information should be obtained from a product design database, taking the form of well-documented product specifications or a Computer-Aided Design (CAD) model. Unfortunately, this information is usually unavailable by the time EOL products are returned. Therefore, the disassembly process needs to deal with incomplete product information, some of which is only revealed during the disassembly process. The challenge is to develop a disassembly plan that is general enough to deal with these uncertainties [5].

Difficulty of operations

Kroll et al. [6] define the term *disassemblability* to quantify the ease of product disassembly. A product is assessed for disassemblability according to the difficulty of the disassembly operation, by assessing it against five major criteria: (a) *component accessibility*, (b) *precision in locating the component*, (c) *force required to perform tasks*, (d) *additional time*, and (e) *special problems* that cannot be categorised in the other areas. Mok et al. [7] summarise the characteristics of an ease of product disassembly as follows:

- Minimal force exertion;
- Quick operation without excessive manual labour;
- Simple mechanism of disassembly;
- Minimal use of tool: ideal disassembly should be performed without tools;
- Minimal part repetition: parts easy to identify at each state of disassembly;
- Easy recognition of fasteners;
- Simple product structure; and,
- Avoidance in usage of toxic material.

Gupta and McLean [8] state that the development of optimal disassembly plans relies on four key phases: (a) *product analysis*, (b) *assembly analysis*, (c) *usage mode and effect analysis* and (d) *dismantling strategy*. Firstly, the product must be analysed and represented systematically. Options regarding the disassembly process can be generated from or represented using the product structure. The process can be considered at two levels, which are the *sequence plan* and the *operation*. The completeness of disassembly is considered a part of the sequence plan.

In summary, the uncertainties and variations found within returned products leads to uncertainties in the disassembly process. Uncertainties and variations of the disassembly process are summarised in Table 2.1.

Table 2.1 Summary of variations and uncertainties in the disassembly process

Category	Specific issues
Uncertainty in EOL condition	• Modification of product during usage phase
	• Condition of product
	• Condition of main component
	• Condition of connective component
Diversity of the supplied products	• Main product structure
	• Physical appearance of components
	• Quantity of components
	• Location of components
	• Variation in manufacture (quality)
Complexity in process planning and operations	• Disassembly sequence plan
	• Disassembly operation plan (considering prior actions)
	• Disassembly process parameters
	• Capabilities and limitations of detection technique
	• Precision of robot's sensors—actuators
External factors	• Technology and design changes
	• Market driven factors

2.1.2 Representation of Product Structure

The structure of a product consists of (a) *components* and (b) *connections* [9]. A *component* is an element that keeps its extrinsic properties, i.e. functionality and material properties, after being detached from the product. A component cannot be further dismantled without using destructive disassembly methods. The *connection* or liaison is a relation that physically connects two components to restrict the motion between them. The task of disassembly is to disestablish these relations in order to separate the relevant components.

Fasteners

A fastener is a component or design element that serves the purpose of connecting other (main) components. Fasteners that are insignificant to the goal of disassembly may be modelled separately to the main components. Lambert and Gupta [9] considers such fasteners as *quasi-components*, which can be discrete components (e.g. screws, rivet, cable, etc.) or part of the main component (e.g. snap-fits). Connection-establishing elements, such as solder and weld joints, that do not form a component in themselves, can be considered *virtual components*.

Product structure

The structure of a product can be represented in many ways, of which two will be detailed here: the *connection diagram* and the *disassembly matrix*.

First, the *connection diagram* (liaison diagram) graphically represents the complete product structure using an undirected graph. The components are represented by nodes and the connections by arcs. According to the level of detail, the graph can be shown in three different forms: (a) *extended form* (b) *reduced form* and (c) *minimal form* (see Fig. 2.1).

In Fig. 2.1a, the product is a composition of three main components—A, B, and C. A and B are connected by mating. B and C are connected with a screw E which is considered a quasi-component. C and A are connected by a weld D which is a virtual component. The *extended form* shows full details of the product with every component and fastener. All fasteners, including virtual connections, are modelled (see Fig. 2.1b). The *reduced form* represents the structure more concisely by hiding the virtual components and using dashed lines for quasi-components. In this case,

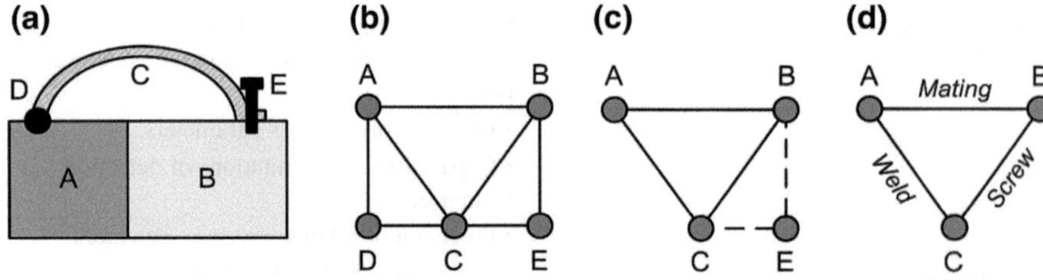

Fig. 2.1 Connection diagram [9]. **a** Product assembly. **b** Extended form. **c** Reduced form. **d** Minimal form

the connections associated with the virtual component, D–A and D–C, are removed. As a result, only connection A–C is retained representing the weld (see Fig. 2.1c). The *minimal form* shows the structure of the product in the most compact way by hiding both virtual and quasi-components. This form represents the product in the simplest way while preserving the information regarding the main components (see Fig. 2.1d).

Second, the product structure can be represented by a *disassembly matrix*, which a computing approach (e.g. Linear Programming (LP) or Integer Programming (IP)) can be used to solve the disassembly planning problem. The disassembly matrix is an $N \times N$ connectivity matrix where N is the number of the components. Each element of the matrix represents the existence of a connection between two corresponding components: "1" if a connection exists, and "0" if it does not. This information is completely represented by the lower left part of the matrix, since the matrix is symmetric and the elements on the diagonal axis are non-applicable. From this matrix, it is clear that the maximum number of connections is $\frac{1}{2}(N)(N-1)$. The disassembly matrix of the example product in Fig. 2.1a is shown in Eq. (2.1).

$$\text{Disassembly Matrix} = \begin{array}{c|ccccc} & A & B & C & D & E \\ \hline A & & & & & \\ B & 1 & & & & \\ C & 1 & 1 & & & \\ D & 1 & 0 & 1 & & \\ E & 0 & 1 & 1 & 0 & \end{array} \tag{2.1}$$

2.1.3 Disassembly Process Representation

The steps of the product disassembly process and their corresponding relationships can be schematically represented in many ways. Lambert and Gupta (2005) [9] summarise these approaches as follows:

Disassembly precedence graph
The disassembly precedence graph expresses sub-tasks of the disassembly process connected and constrained by precedence relationships. This can be represented in two forms: as a *component-oriented* or *task-oriented* graph (see Fig. 2.2). The arrows communicate the ordering in which tasks must be performed. This technique was originally used for assembly process representation and assembly line-balancing problems. Gungor and Gupta [10] introduce this to the disassembly process due to its simplicity. However, a major disadvantage is that a complete disassembly sequence cannot be expressed in one graph [11].

Disassembly tree
The disassembly tree expresses all possible choices for disassembly sequences, and is derived from a table containing all possible sequences sorted by level and

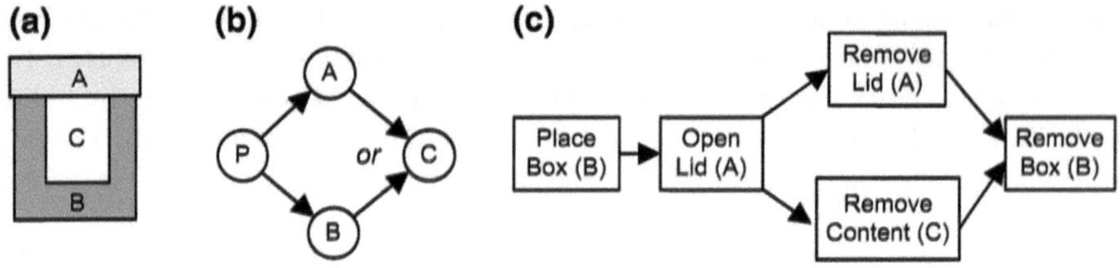

Fig. 2.2 Disassembly precedence [9]. **a** Assembly. **b** Component-oriented. **c** Task-oriented

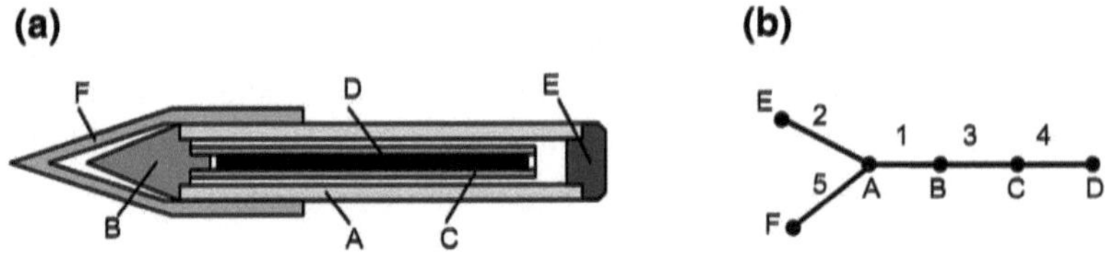

Fig. 2.3 Example product Bourjault's ballpoint [9]. **a** Assembly. **b** Connection diagram

operation type. A widely-used example is the *Bourjault tree* [12]. Two major drawbacks are the complexity arising in complex products and difficulty in representing parallel operations. Figure 2.4 shows a Bourjault tree representation of the disassembly process of a sample product, the Bourjault's ballpoint, which is shown in Fig. 2.3. This product will also be used to demonstrate the representation methods described in the following sections.

State diagram

The state diagram represents the disassembly sequence as an undirected graph, where each node represents a state of disassembly. This can be categorised into two approaches: (a) *connection-oriented* [13] and (b) *component-oriented* [14, 15] (see Fig. 2.5). All possible combinations of connections are represented by the nodes. Each edge represents the establishment or disestablishment of a connection. The major advantages are that the disassembly sequence of the complete product can be demonstrated in one diagram, and the diagram is compact even for complex products. However, state diagrams are unable to show how the disestablishment of some connections cannot be done individually without affecting a combination of related connections.

Kara et al. [16] used a connection-oriented state diagram representation to develop a graphical representation method, the *disassembly-sequence diagram*, for representing the disassembly sequence to and from different stages of the process for selective disassembly. This diagram can be automatically generated from the liaison and precedence relations. An example is shown in Fig. 2.6.

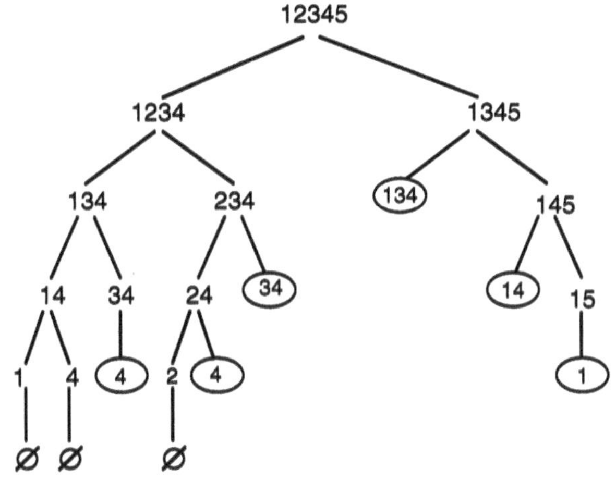

Fig. 2.4 Disassembly tree of the Bourjault's ballpoint [9]

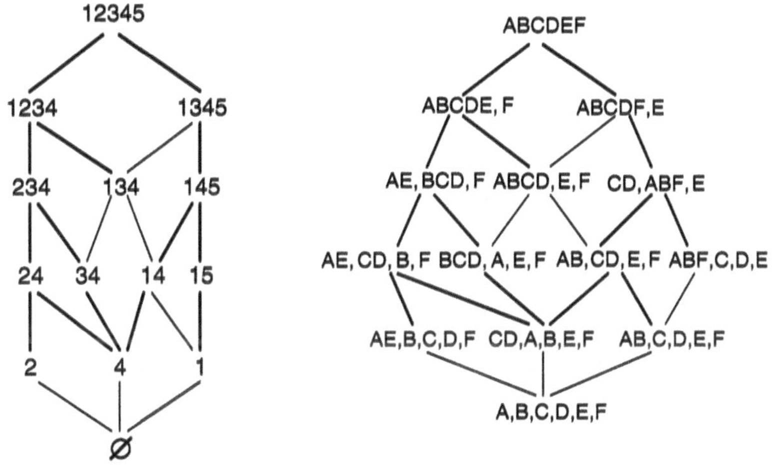

Fig. 2.5 State diagram of the Bourjault's ballpoint [9]

AND/OR graph (Hypergraph)

This graph represents disassembly sequences based on subassemblies. A process is represented by multiple-arcs (hyper-arcs) pointing from a parent to its child components (subassemblies) (see Fig. 2.7). This overcomes the drawback of the state diagram. However, a major drawback is the complexity of the visual representation, which may become difficult to read when the number of components increases. Lambert [17] proposes a simplified version of this graph named the *concise AND/OR graph*. Further developments, aimed at representing the product model and its constraints more accurately, include the arborescence with hypergraph [18], Petri net [19], and Hybrid graphs [20].

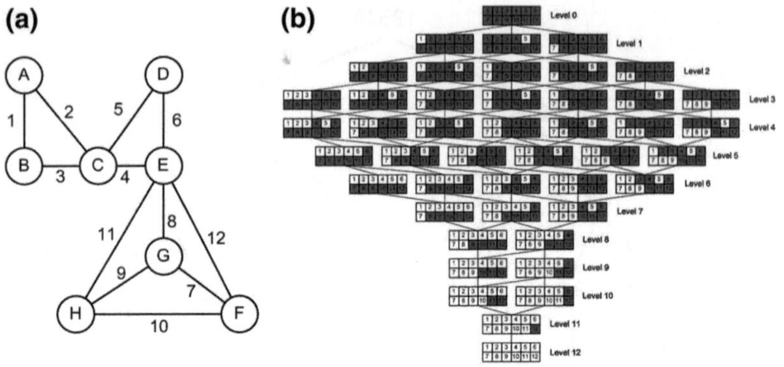

Fig. 2.6 Disassembly-sequence diagram [16]. **a** Liaison diagram. **b** Diassembly-sequence diagram

Fig. 2.7 AND/OR graph of the Bourjault's ballpoint [9]

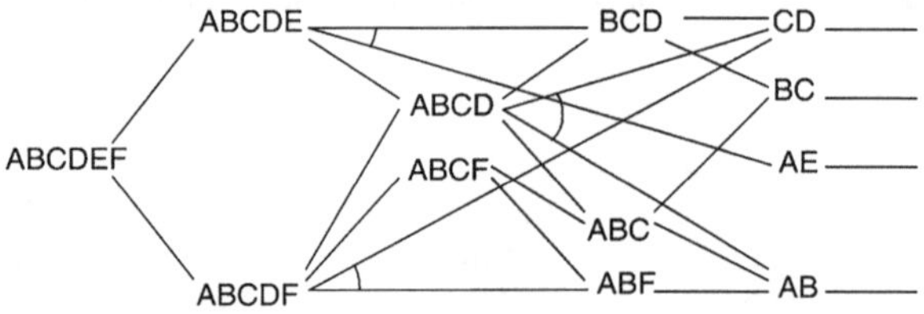

2.1.4 *Disassembly Sequence Planning (DSP)*

A disassembly sequence is a procedure for the disestablishment of connections and detachment of parts in the disassembly operation. The initial state is defined as the complete product, and the final state, as a state where all desired components or subassemblies have been separated. The main purpose of disassembly sequence planning (DSP) is to find the optimal sequences of disassembling products with respect to certain factors, e.g. cost-effectiveness, material return, component recovery, and duration of operations. Theoretically, the number of possible sequences increases exponentially according to the number of components. Therefore, finding the optimal solution is considered an NP-complete optimisation problem [4].

Lambert [5] summarises effective methodologies based on a product-oriented approach as follows. As adaptability is required for a flexible automatic disassembly system, the main theme of this book, emphasis is placed on the adaptive planners.

Mathematical programming (MP) method

The mathematical programming (MP) method aims to make the internal variables converge to their optimum value without considering the complete search space. The problem model is derived from a hypergraph (AND/OR graph). Costs are assigned to each action (arc) with respect to subassembly components (i.e. parent and child) and stored in a transition matrix. This can then be effectively solved by mathematical solvers, e.g. using Linear Programming (LP), Mixed Integer Programming (MIP), or Dynamic Linear Programming (DLP). Petri nets are also used in case of a dynamic approach.

Heuristic methods
Gungor and Gupta [21] present a heuristic algorithm used to find near-optimal solutions to the disassembly sequencing problem. Near-optimal solutions are considered instead of optimal solutions, which are sometimes difficult to find due to the size of the search space. This method requires information of the precedence relationship among each of the components and the difficulty in performing each action. Efficiency is evaluated by the authors based on disassembly time. A case study regarding the DSP of a cell phone using the heuristic method and different search algorithms, e.g. greedy k-best and A*, is examined by Lambert and Gupta [22].

Artificial intelligence (AI) methods
Various techniques are used in artificial intelligence to generate and utilise constraints and reduce the size of the search space. Lambert [5] reviews typical AI techniques for disassembly sequence planning, including simulated annealing algorithms, genetic algorithms (GA), fuzzy sets, neural networks, multi-agent systems, and Bayesian networks. Other novel algorithms that have been efficiently applied to DSP include ant-colony optimisation [23], case-based reasoning [24] and rule-based sequence generation on clustering graphs [25].

Adaptive planner
An adaptive planner generates a disassembly sequence with respect to the uncertainties and unexpected circumstances encountered during the disassembly operation. Due to its particular relevance to automated disassembly, a number of publications relating to adaptive planners have been reviewed in this section. The literature handles the problem at two levels: the (a) *process planning level* and (b) *operation level.*

In the *process planning level*, Tang [26] proposes using a Fuzzy Petri net to model the dynamics of disassembly, including the uncertainties in product condition and human factors. The system is trained with data and feedback from the actual disassembly, and selects the appropriate disassembly plan based on past experience. Turowski et al. [27] presents an implementation of a Fuzzy Coloured Petri Net for balancing a disassembly line. Grochowski and Tang [28] propose a learning approach using a Disassembly Petri Net (DPN) and Hybrid Bayesian network. Veerakamolmal and Gupta [29] propose using case-based reasoning (CBR) to generate disassembly plans for multiple products. The plan for a new product is adapted from existing plans by deriving it from a base case. Gao et al. [30] propose using a Fuzzy Reasoning Petri Net to adaptively generate the disassembly sequence according to the condition of the product observed at each state. Decisions are made based on the estimated value returned, hazard level, and disassembly cost.

In the *operation level*, Salomonski and Zussman [31] propose using a predictive model with DPN to adaptively generate the disassembly process plan according to real-time measurements conducted by a robot arm. Lee and Bailey-Van Kuren [32] address the uncertainties in the operation level by automatically recovering from a visually-detected error. In addition, Martinez et al. [18] propose a dynamic sequence generation method that generates an optimal disassembly plan during operations in response to unpredictable situations, e.g. failure to remove a corroded part, replacement of screws, etc. This system is modelled and controlled by a *multi-agent system* (MAS). ElSayed et al. [33] use GA to generate an optimal disassembly sequence according a supplied bill-of-materials (BOM) and components detected in real time. Relations defined in the original BOM must be preserved.

In conclusion, the existing adaptive planners deal with many types of uncertainty experienced during the disassembly process. The uncertainties relate to variations in the component conditions that deviate from the ideal case. The ability for an adaptive planner to handle these uncertainties stems from its ability to appropriately adapt existing knowledge into a new plan according to sensed information. Machine learning techniques are used to allow the system to improve its performance from past experience. However, the structure of the product, e.g. BOM and CAD model, generally needs to be supplied a priori. A methodology accounting for an uncertainty in the general product structure in real-time has not yet been proposed in any research. In addition, the learning process has only been implemented in the planning level. Hence, learning at the operation level, such as in optimising process parameters, should be further investigated.

2.2 Completeness of Disassembly

The completeness of disassembly can be categorised into two types: (a) *complete disassembly* and (b) *incomplete disassembly*. A *complete* or *full disassembly* is the process that separates every single component of the product. This is rarely done due to the technical constraints (particularly the complexity and the uncertainties in the operation) and high labour costs. On the other hand, the *incomplete* or *selective disassembly* separates only the desired components or subassemblies, and terminates when the desired depth of disassembly is reached. Disassembly becomes more cost-efficient with a strategic choice of disassembly targets. Reasons for selective disassembly include recovering modules or components for use as spare parts, separating those that contain hazardous substances, and improving the quality and quantity of shredder residue [17].

Figure 2.8 illustrates the situation in maximising the profit in the disassembly process. The disassembly range refers to the completeness of disassembly. The cost of disassembly is due to operation time, varying according to the number and type of connections to be disestablished. This increases with the desired completeness of disassembly. Disassembly is economically feasible when the total profit from treating or recycling all products exceeds the cost of disassembly. The optimal strategy is the point at which the maximum profit can be obtained [34, 35].

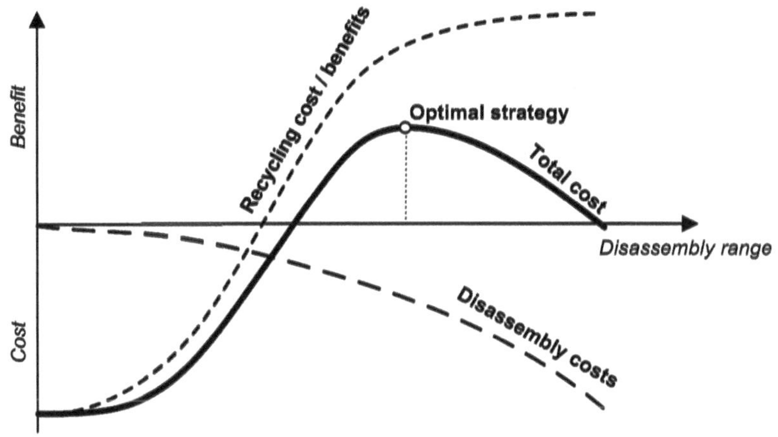

Fig. 2.8 Determination of optimal disassembly strategy [34]

The outcome of selective disassembly can be one of the three following types [9].

- **Homogeneous components**: parts that cannot be physically disassembled.
- **Complex components**: components comprised of a number of homogeneous components, joined together with fasteners, which can only be separated using destructive disassembly.
- **Modules**: sets of components that perform a self-contained function. Modules can be further disassembled via non-destructive or semi-destructive operations. However, maintaining their original condition and functionality can allow reuse of the entire module.

The researchers currently focus on developing a methodology to find optimal disassembly sequences in which the completeness of disassembly is taken into account. Kara et al. [36] propose the methodology of developing the optimal selective disassembly sequence which is the reverse of the methodology for assembly presented by Nevins and Whitney [37]. The disassembly sequences are generated from the product specifications, namely list of parts and subassembly, precedence rules, product representation model, and disassembly sequence diagram. Subsequently, the optimal sequences for removing the selected parts are obtained by removing invalid sequences according to liaison analysis. In regard to this concept, software that automatically generates and visualises optimal sequences of selective disassembly from specified constraints is developed by Kara et al. [16, 38].

2.3 Disassembly Operations

2.3.1 Types of Fasteners

The disassembly operation is divided into two main tasks: *disestablishing fasteners* and *detaching main components*. The main component is detachable if the associated

Table 2.2 Disassembly methods according to fastener type

Discrete components	Not deformed	Bundling	Shear cut
		Spring	Deform/pull
		Screw, bolt, nut, washer	Unscrew, drill
	Reversibly deformed	Cotter pin, staple	Pull
	Irreversibly deformed	Rivet	Pry out, drill
		Adhesive: glue, seal	Peel, pry out, break
Parts of components	Reversible connection (semi-reversible)	Surface: mating	Remove
		Surface: press fit	Pull, pry out
		Snap fit	Deform, pry out, pull
	Irreversible connection	Surface: press fit	Pull, pry out
		Surface: mould	Break
		Seam fold	Deform
		Seal	Peel, pry out, break
Virtual components	Irreversible	Solder	Shear cut, break, melt
		Weld	Saw cut, break

connections are located and disestablished. Specific techniques are required for effectively disestablishing different types of fasteners. Lambert and Gupta (2005) [9] categorise the fasteners commonly found in mechanical and electronic-electrical products into 13 types. Different types of fasteners require different disestablishment methods and display different levels of reversibility. A summary of fastener types and their respective disassembly methods is shown in Table 2.2.

2.3.2 Dismantling Techniques

Disassembly operations can be broadly categorised into three types: (a) *non-destructive*, (b) *semi-destructive*, and (c) *destructive* disassembly. The characteristics of each category are explained in detail as follows.

Non-destructive disassembly
All outputs of non-destructive disassembly remain undamaged. This is desired for maintenance, component reuse and remanufacture. All fasteners within the product must be reversible or semi-reversible. The dismantling of reversible fasteners (e.g. screws) is generally easier than that of semi-reversible fasteners (e.g. snap-fits). The operation cost is generally high, as high flexibility is required, particularly due to difficulties such as rust and partial damage. Even though a number of tools have been specially developed to facilitate non-destructive disassembly, e.g. for the disassembly of screws [39] and snap-fits [40], the non-destructive approach is still generally economically infeasible [3].

Semi-destructive disassembly

The semi-destructive approach aims to destroy only connective components, e.g. via breaking, folding or cutting, leaving main components with little or no damage. This increases the efficiency of the operation and has been found in many cases to be economically feasible. Many research works relating to automatic disassembly use semi-destructive techniques to overcome the uncertainties in the product condition and geometry. Examples of such techniques include the drilling out of screw heads during the disassembly of electric motors [41], creation of new surfaces allowing torque transmission for unscrewing [34] and cutting off of screw heads using a cut-off wheel [42].

Destructive disassembly

Destructive disassembly deals with the partial or complete destruction of obstructing components. Components or irreversible fasteners, e.g. welds, are destroyed using destructive tools such as a hammer, crowbar or grinder. These operations are fast, efficient and inherently flexible. As a result, destructive disassembly is economically feasible and commonly performed in industry practice. One common application of destructive disassembly is in the opening of a covering component to reach the more valuable components inside. Examples include the breaking of the separating line [34] and using plasma arc cutting to destroy the metal casing of consumer appliances [43].

In summary, semi-destructive and destructive disassembly allow techniques that are more capable of efficiently dealing with the uncertainties in product condition, therefore allowing more economically feasible operation. On the contrary, non-destructive disassembly tends to have high operation costs but may be unavoidable in maintenance or for component reuse. More detail regarding the operations and specially-developed tools can be found in Sect. 3.3.2.

2.4 Conclusions

Disassembly can be a key step in an efficient EOL treatment process, however, is usually economically infeasible due to high operating costs relating to the variation and uncertainties in the products and process, as summarised in Table 2.1. This chapter presents three major considerations which should be addressed to improve the economic feasibility of disassembly.

Firstly, the disassembly plan can be optimised with respect to a goal, which can be operating time or cost. A number of techniques regarding the representation of the product structure and disassembly process are described. With an appropriate representation, an optimal or near-optimal ordering of the disassembly operations can be found via various optimisation strategies described in Sect. 2.1.4. Particularly the adaptive planners are of interest, since they are able to respond to minor uncertainties like product damage. A significant amount of product knowledge, e.g. the product structure or a CAD model, are required before planning.

Secondly, regarding the completeness of disassembly, performing selective disassembly to a certain depth is more feasible than the full disassembly. The optimal disassembly depth should be determined during the planning phase.

Finally, the difficulty of the disassembly operations results from the type of fasteners used, and the product and fasteners' conditions. Different fasteners can be disestablished using different tools and techniques. Semi-destructive and destructive operations are generally preferable due to the shorter operating time and effective operation in spite of uncertainties.

In conclusion, the main source of difficulty in the disassembly process is the need to deal with a high level of uncertainties and variations. If the disassembly is not conducted by the product manufacturers, information regarding the product is generally at first incomplete. Even when the expected outcomes are known, poor product or fastener condition may require deviations to the usual plan. This causes higher operating time in manual disassembly, and is a primary factor hindering the industrial application of automatic disassembly.

References

1. Boothroyd G, Alting L (1992) Design for assembly and disassembly. CIRP Ann—Manuf Technol 41(2):625–636
2. Chiodo J (2005) Design for disassembly guidelines. Active Disassembly Res. http://www.activedisassembly.com/downloads/ADR_050202_DFD-guidelines.pdf
3. Duflou JR, Seliger G, Kara S, Umeda Y, Ometto A, Willems B (2008) Efficiency and feasibility of product disassembly: a case-based study. CIRP Ann—Manuf Technol 57(2):583–600
4. Gungor A, Gupta SM (1998) Disassembly sequence planning for products with defective parts in product recovery. Comput Ind Eng 35(1–4):161–164
5. Lambert AJD (2003) Disassembly sequencing: a survey. Int J Prod Res 41(16):3721–3759
6. Kroll E, Beardsley B, Parulian A (1996) A methodology to evaluate ease of disassembly for product recycling. IIE Trans (Institute of Industrial Engineers) 28(10):837–845
7. Mok HS, Kim HJ, Moon KS (1997) Disassemblability of mechanical parts in automobiles for recycling. Comput Ind Eng 33(3–4):621–624
8. Gupta M, McLean CR (1996) Disassembly of products. Paper presented at the 19th international conference on computers and industrial and engineering, Computer Industrial Engineering
9. Lambert AJD, Gupta M (2005) Disassembly modeling for assembly, maintenance, reuse, and recycling. CRC Press, Boca Raton, Fla.
10. Gungor A, Gupta SM (2002) Disassembly line in product recovery. Ann Rev Control 40(11):2567–2589
11. Tumkor S, Şenol G (2007) Disassembly precedence graph generation. In: SAM 2007—IEEE international symposium on assembly and manufacturing, pp 70–75
12. Bourjault A (1984) Contribution à une approche méthodologique de l'assemblage automatisé: elaboration automatique des séquences opératoires (Contribution to a systematic approach of automatic assembly: The automatic generation of operation sequences). Université de Franche Comté, Besançon, France
13. Fazio D, Whitney TL (1987) Simplified generation of all mechanical assembly sequences. IEEE J Robot Autom RA-3 (6):640–658
14. Homem De Mello LS, Sanderson AC (1990) AND/OR graph representation of assembly plans. IEEE Trans Robot Autom 6(2):188–189

15. Woller JD (1992) A combinatorial analysis of enumerative data structures for assembly planning. J Des Manuf 2(2):93–104
16. Kara S, Pornprasitpol P, Kaebernick H (2006) Selective disassembly sequencing: a methodology for the disassembly of end-of-life products. CIRP Ann—Manuf Technol 55(1):37–40
17. Lambert AJD (1999) Linear programming in disassembly/clustering sequence generation. Comput Ind Eng 36(4):723–738
18. Martinez M, Pham V-H, Favrel J (1997) Dynamic generation of disassembly sequences. In: IEEE symposium on emerging technologies and factory automation, ETFA, pp 177–182
19. Zussman E, Zhou M, Caudill R (1998) Disassembly Petri net approach to modeling and planning disassembly processes of electronic products. In: IEEE international symposium on electronics and the environment, pp. 331–336
20. Wang Y, Li F, Li J, Chen J, Jiang F, Wang W (2006) Hybrid graph disassembly model and sequence planning for product maintenance. Paper presented at the IET conference publications
21. Gungor A, Gupta SM (1997) An evaluation methodology for disassembly processes. Comput Ind Eng 33(1–2):329–332
22. Lambert AJD, Gupta SM (2008) Methods for optimum and near optimum disassembly sequencing. Int J Prod Res 46(11):2845–2865
23. Shan H, Li S, Huang J, Gao Z, Li W (2007) Ant colony optimization algorithm-based disassembly sequence planning. In: IEEE international conference on mechatronics and automation
24. Shih L-H, Chang Y-S, Lin Y-T (2006) Intelligent evaluation approach for electronic product recycling via case-based reasoning. Adv Eng Inform 20(2):137–145
25. Kaebernick H, O'Shea B, Grewal SS (2000) A method for sequencing the disassembly of products. CIRP Ann—Manuf Technol 49(1):13–16
26. Tang Y (2009) Learning-based disassembly process planner for uncertainty management. IEEE Trans Syst Man and Cybern—Part A: Syst Humans 39(1):134–143
27. Turowski M, Tang Y, Morgan M (2005) Analysis of an adaptive fuzzy system for disassembly process planning. In: IEEE international symposium on electronics and the environment, pp 249–254
28. Grochowski DE, Tang Y (2009) A machine learning approach for optimal disassembly planning. Int J Comput Integr Manuf 22(4):374–383
29. Veerakamolmal P, Gupta SM (2002) A case-based reasoning approach for automating disassembly process planning. J Intell Manuf 13(1):47–60
30. Gao M, Zhou MC, Tang Y (2005) Intelligent decision making in disassembly process based on fuzzy reasoning Petri nets. IEEE Trans Syst Man Cybern B Cybern 34(5):2029–2034
31. Salomonski N, Zussman E (1999) On-line predictive model for disassembly process planning adaptation. Robot Comput-Integr Manuf 5(3):211–220
32. Lee K-M, Bailey-Van Kuren MM (2000) Modeling and supervisory control of a disassembly automation workcell based on blocking topology. IEEE Trans Robot Autom 16(1):67–77
33. ElSayed A, Kongar E, Gupta SM, Sobh T (2012) A robotic-driven disassembly sequence generator for end-of-life electronic products. J Int Robot Syst: Theory Appl 68(1):43–52
34. Feldmann K, Trautner S, Meedt O (1996) Innovative disassembly strategies based on flexible partial destructive tools. Ann Rev Control 23:159–164
35. Desai A, Mital A (2003) Evaluation of disassemblability to enable design for disassembly in mass production. Int J Ind Ergon 32(4):265–281
36. Kara S, Pornprasitpol P, Kaebernick H (2005) A selective disassembly methodology for end-of-life products. Assembly Autom 25(2):124–134
37. Nevins JL, Whitney DE (1989) Concurrent design of products and processes: a strategy for the next generation in manufacturing. McGraw-Hill, New York
38. Pornprasitpol P (2006) Selective disassembly for re-use of industrial products. University of New South Wales
39. Seliger G, Keil T, Rebafka U, Stenzel A (2001) Flexible disassembly tools. In: IEEE international symposium on electronics and the environment, pp 30–35

40. Braunschweig A (2004) Automatic disassembly of snap-in joints in electro-mechanical devices. In: the 4th international congress mechanical engineering technologies'04, Varna
41. Karlsson B, Järrhed J-O (2000) Recycling of electrical motors by automatic disassembly. Meas Sci Technol 11(4):350–357
42. Reap J, Bras B (2002) Design for disassembly and the value of robotic semi-destructive disassembly. In: ASME design engineering technical conference, pp 275–281
43. Uhlmann E, Spur G, Elbing F (2001) Development of flexible automatic disassembly processes and cleaning technologies for the recycling of consumer goods. In: IEEE international symposium on assembly and task planning, pp 442–446

Chapter 3
Disassembly Automation

Abstract Disassembly is one of the key steps in the efficient treatment of End-Of-Life products, however has generally been neglected in industry practice due to cost constraints. This is because disassembly has traditionally been restricted to manual labour. Automation assisting or substituting for human workers may be a lower-cost option. However, the associated technical challenges need to first be addressed. This chapter presents the principle of disassembly automation, as well as the basic elements needed in a disassembly system. A number of research works in regard to automated disassembly systems and innovative disassembly tools are also described.

3.1 Introduction

Nowadays, automation plays an important role in the modern manufacturing industry. This automation was introduced to the manufacturing industry almost a century ago, with the introduction of feedback control [1, 2]. Automation has in general proved to be more cost-effective than human workers. A major advantage of automation is the ability to handle repetitive tasks quickly with high precision and accuracy. In addition, robots can be applied to hazardous tasks and in working environments that are undesirable for the human workers due to e.g. contamination or radiation [3]. The skilled labour required to achieve the high accuracies, or the measures required to meet health and safety requirements for human workers lead to high operating costs, particularly in developed countries. The primary advantages that automation has given to the manufacturing process can be summarised as: (a) cost effectiveness, (b) efficiency and accuracy, and (c) the ability to distance human workers from hazardous activities.

Many similar issues are faced in the disassembly process. Economical infeasibility is a major reason behind disassembly being generally neglected by the EOL industry [4]. It is difficult to economically justify the implementation of manual disassembly due to the generally-low economic returns from EOL treatment. The costs are high,

S. Vongbunyong and W.H. Chen, *Disassembly Automation*, Sustainable Production, Life Cycle Engineering and Management, DOI 10.1007/978-3-319-15183-0_3

as the process is generally labour-intensive, and is often met with complications due to the variations and uncertainties in EOL products returned. The process may also be hazardous depending on the subject of disassembly. For example, the disassembly of electric vehicle batteries deals with risks due to the chemical composition and any charge remaining in the cells [5]. Automation has a high potential to reduce these problems in the disassembly process [6]. A number of attempts have been made to implement automation in the disassembly process. However, apart from manufacturers' own efforts to remanufacture their own products, the latest developments are still in the research phase and have not yet been implemented at the industrial level. The problem relates not only to the technical but also the economic aspects. Scharke [7] summarises the challenges of automated disassembly as follows:

- the variety of manufacturers and different product types,
- variety in design and product structure,
- small lot sizes and insufficient collection logistic,
- varying product condition after the usage phase,
- changes in standard components,
- insufficient disassembly tools,
- changes in legislation, and
- changes in market demand and prices.

The uncertainties due to the variations in returned products make the disassembly process and facility hard to design and economically justify. This makes the process plans and operations more complicated than those required for assembly. The variety in product condition may occur due to component replacement, upgrading/downgrading, and damage during the usage phase. In general, an optimised disassembly process can be planned according to sufficient product-specific information, e.g. computer aided design (CAD) model, bill-of-materials (BOM), or precedence relations. between disassembly operations. Unfortunately, this knowledge is only available for a small subsection of products at the EOL stage.

For disassembly, manual labour has traditionally been preferred over automation due to the human's advantages in (a) *perception*, (b) *dexterity*, and (c) *intelligence*. These aspects allow a human to interact flexibly with its environment. *Perception* refers to the sensor systems ability to perceive information about the process. *Dexterity* relates to the proficiency of carrying out the physical operation. *Intelligence* relates to the ability to plan and control the process according the knowledge of the product. Unlike the assembly process, information about the product is often inaccurate or incomplete at beginning of the process, but is revealed during the process [8]. The successful implementation of disassembly automation requires significant improvement in these attributes, which may be achieved with the aid of the human worker.

In practice, to limit the variations encountered in the process, each system is designed for disassembling a particular product type or family. Therefore, the technical requirements of the dismantling tools, relating to e.g. the type of techniques, product size and weight, can be limited. A specifically-designed system can satisfy the cost constraint given a sufficiently-large lot size. However, the lot size of returned EOL products is unpredictable and may lead to difficulty at the strategic planning level. In addition, the profit depends on the market value of the recycled and reused parts, which changes according to demand and legislation. Therefore, strategic planning regarding these aspects should be done before developing the automated system.

In summary, automation has the potential to increase the economic feasibility of the disassembly process, in reducing operating costs and risk to the human worker. However, the technical requirements must first be met: the system needs to be flexible and robust enough to deal with uncertainties in the product and process. Due to the high level of intelligence, perception, and dexterity required in the system, humans may still need to play specific roles, e.g. in supervision, high-level planning, or performing sophisticated tasks.

This chapter explains the concept of disassembly automation and state-of-the-art developments within the field. An overview of disassembly automation and the integration of the various components are explained in Sect. 3.2. An overview of aspects relating to mechanical design is given in Sect. 3.3. Finally, existing developments in disassembly automation are explained in Sects. 3.4 and 3.5. It is noted that the aspects relating to perception are described in Chap. 4 and cognitive robotics, the artificial intelligence technique used in this research, is explained in Chap. 5.

3.2 Principle of Disassembly Automation

Disassembly automation may refer to a single disassembly cell that is able to complete the key steps of the disassembly process, such as dismantling and separation, or a system comprised of many (potentially more specialised) cells. Configuring a disassembly system as a *modular system* is recommended by [9] to fulfil the economic and technical requirements, since a modular system allows the flexibility of future adaptations to other products with little modification in the hardware and software. Modules may be added, removed, or modified in order to suit the new products. Products that have certain similar features, e.g. size and connection methods, may be grouped into *disassembly families*, as proposed by [10]. As the same operations are required to disassemble products in the same family, the same disassembly tools can be used. This method reduces the technical requirements of the modules needed for the disassembly system. In regard to a modular system, [6] summarise the basic modules as follows:

- industrial robots or manipulating devices;
- disassembly tools specially designed for robots and tasks;
- gripping devices;
- feeding systems of the supply products;
- transport systems;
- fixture systems;
- storage systems for parts and tool;
- manual disassembly stations;

- vision systems;
- sensor systems;
- intelligent control units; and
- product databases.

This comprehensive summary can be taken as a guide for the modules required in the disassembly system. Not all are necessary in every system, and should be adapted on a case-to-case basis.

Excluding any manual input to the process, the development of a disassembly automation system can be considered in three broad subsystems: (a) *mechanical design and control*, (b) *vision and sensors*, and (c) *planning and artificial intelligence*. For this research, these subsystems are implemented as distinct *operating modules*. Figure 3.1 illustrates the implementation and interconnections between the operating modules. An overview of the uncertainties and variations expected to be addressed by each of these modules is shown in Table 3.1.

Mechanical Design and Control

The disassembly system physically interacts with the products via its mechanical units. The mechanical design and control of the system focuses on the physics of these units and includes the machine design, kinetics, and kinematics of the units. Motion controllers control the system at the sensor-actuator level. Theory in the *control of dynamic systems* or *feedback control* [11] should be considered if the low-level development of these components is necessary.

A cell for disassembly automation consists of a number of mechanical components. The main component is a *robot* for the manipulation of disassembly tools. Various kinds of *disassembly tools* can be used, depending on the connections to be disestablished. These tools can be traditional hand or power tools, e.g. screwdriver, drill or grinder, or innovative disassembly tools designed specifically for the efficient disestablishment of connectors (e.g. snap-fits [12]). After the connectors have been disestablished, a *gripping device* suitable for the various geometry and dimensions of the parts is used to remove detached parts. A *tool changer* may be needed if the system requires multiple disassembly tools and/or grippers for the one manipulator. Products need to be fixed to the system via a *fixture* that is capable of dealing with the various geometry and dimensions of the products. Other mechanical supporting

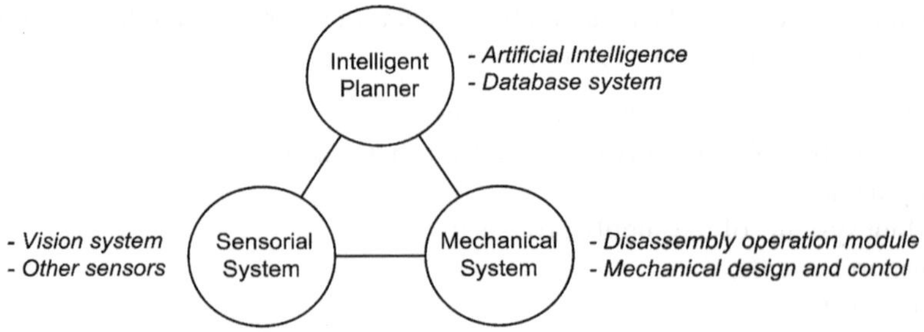

Fig. 3.1 Overview of the system

Table 3.1 Uncertainties and variations addressed by the three operating modules

Primary uncertainty	Specific issue	Related operating module		
		Intelligent agent	Vision system	Disassembly operation unit
Uncertainty in EOL condition	Modified product	•		
	EOL condition of product	•		
	EOL condition of main component	•		
	EOL condition of connective component	•		
Diversity of the supplied products	Main product structure	•		
	Physical appearance of the components		•	
	Quantity of the components	•	•	
	Location of the components		•	
	Variation in manufacture (quality)	•	•	•
Process and operations	Disassembly sequence plan	•		
	Disassembly operation plan (prior actions)	•		
	Disassembly process parameters	•		
	Capability of detection technique	•	•	•
	Precision of robot's sensors—actuators	•	•	•
	Wear of disassembly tool	•		•
External factor	Technology and design change	n/a		
	Market driven factor			

components deal with the logistics of the system, including transportation, feeding, and storage of the products. *Transportation* is generally performed using conveyor systems, which can be designed to suit the configuration of the workstations.

Vision System and Sensors

Due to the uncertainties and variations in the supply, information about a product to be disassembled is generally at first incomplete, and revealed during the disassembly process. Sensor systems are used to perceive this information during the process. A sensor system capable of sensing the scene at a distance, e.g. using cameras or distance sensors, is generally used to obtain general information regarding the existence and location of the product and components. Other sensors, such as force and torque sensors, support the system by providing additional information regarding specific tasks [6]. The sensed information can be used at the planning and operational levels to adapt the process accordingly.

Planning and Artificial Intelligence

The planning and artificial intelligence subsystem is responsible for managing the data flow within the system, controlling the physical process to appropriately respond to the information received from the sensors and prior knowledge. This can take the form of an *intelligent agent*. The process needs to be carried out according to a disassembly sequence plan (DSP) and disassembly process plan (DPP), which may be predefined or automatically generated by the agent. A number of artificial intelligence techniques have been applied in automated disassembly systems, e.g. genetic algorithms [13], Petri-nets [14] and cognitive robotics [15]. The technique defines the behaviour of the agent controlling the physical system.

Data is abstracted at various levels to reduce the complexity of the planner. In this case, the agent controls the system by supervising other operating modules, which handle details at a lower level. As a result, the desired behaviour of the agent can be better formulated and generalised. The levels of control are a key aspect of the system architecture and should be determined prior to the implementation of this subsystem. As an example, the control layers used in this research are shown in Table 3.2. This system consists of three levels of control: (a) *high-level*, (b) *mid-level*, and (c) *low-level*.

The intelligent agent or planner controls the system at the *high-level*. The agent generates the disassembly sequence plan and process plan. The process plan is a sequence of operations that can be divided into tasks, which can be executed by the respective operating modules. Plans are generated according to the knowledge base and the information sensed during the disassembly process. As a result, the system is able to conduct complex procedures, responding to prior knowledge and sensed information according to specified behaviour.

At the *mid-level*, the intelligent agent controls the information flow in the system. This level concerns detail at the operation level. The agent communicates between the vision and mechanical systems respectively regarding information such as the location of components and desired operations.

Table 3.2 Tasks separated into levels of control and operating modules

Control	Module		
	Intelligent agent	Vision system and sensors	Mechanical system
High-level	–Sequence planning	n/a	n/a
	–Process planning		
	–Task planning		
	–Knowledge base		
	–Behaviour control		
Mid-level	–Information flow	–Object localisation	–Operation procedure
		–Object recognition	–Path planning
		–Signal interpretation	–Trajectory control
Low-level	n/a	–Image pre-processing	–Motion control
		–Camera control	–Force-torque control
		–Image grabber	
		–Data acquisition	
		–Signal processing	

The tasks of the vision system and sensors are to sense the data from the external world, interpret this data, and communicate the relevant abstract information to the intelligent agent upon request. The intelligent agent calls the sensing procedures, e.g. to locate an object and to sense the force exerted on the tool. The *low-level* acquires the relevant data, e.g. a pre-processed image or voltage signal associated with a force, which may be processed using hardware. Algorithms in the *mid-level* then interpret this data to answer the request of the agent.

The tasks in the mechanical system relate to the physical enactment of actions as requested by the agent. The functions in the *mid-level* generate the desired tool trajectory according to the command of the intelligent agent. The desired motion or force-torque control, which concerns actuator control with respect to end-effector position, speed and/or external forces, is then executed at the *low-level.*

The implementation of the operating modules within this research is further explained Sects. 3.5 and 6.4.1.

3.3 Mechanical Design and Control

In a similar way to assembly automation, disassembly automation carries out the process by physically interacting with the products to be disassembled as well as the working environment. Mechanical design concerns the development of tools that are capable of performing the required task. Tools relevant to the disassembly process can be categorised into the following three types: (a) *manipulators,* (b) *disassembly tools,* and (c) *handling devices.*

3.3.1 Manipulator

The manipulator can be a robot arm or other device that facilitates the movement and/or use of various components in the disassembly process, such as disassembly tools, products and sensors. Disassembly systems may consist of a single or multiple manipulators performing different functions and working together. A robot arm is generally selected as the manipulator in disassembly automation due to its versatility. An industrial robot is normally a stand-alone system including a controller sufficient to fully control the robot at the *low-level.* Force-torque control, path generation and basic motion control are typically built-in features of the robot. Physical requirements of the disassembly tasks, e.g. workspace, payload, accuracy and accessibility of the disassembly tool need to be considered when designing or choosing a robot.

Each manipulator needs to have sufficient *degrees of freedom* (DOF) to perform the tasks. In general, 3–7 DOF industrial robots are used according to the task required. 3–4 DOF robots, such as SCARA, Cartesian, and gantry robots, are usually sufficient for pick-and-place tasks. 4–6 DOF robots, e.g. articulated and delta-star robots [3], are typically used for more complex handling of the disassembly tool or object. A 6 DOF robot generally allows the manipulation of an object to an arbitrary position and orientation within the robot's workspace. 7 or more DOF robots may be selected to improve reachability or eliminate problems regarding singularities in movement.

The requirements in *workspace* and *payload* can be determined from the weight and expected forces from the disassembly tools and size of the product to be disassembled. The configuration of the robot should also be selected according to the requirements of the task. Robot configurations can be considered to fall into two main categories: (a) *serial robots* and (b) *parallel robots*.

Serial Robots
Serial robots have been used in industry for many decades and are actuated via links joined in a serial structure. Serial robots generally have a large workspace and high dexterity; however, suffer from a low payload due to the location of the actuators. Examples of serial robots include Selective Compliant Articulated Robot Arms (SCARA), articulated, gantry and Cartesian robots (see Fig. 3.2).

Parallel Robots
On the contrary, parallel robots have only been recently introduced to industry, and display the opposite characteristics. In comparison to serial robots with similar size, weight, and number of actuators, parallel robots operate in a very limited workspace and dexterity. However, due to the high rigidity of the parallel structure, parallel robots can handle considerably larger payload with high accuracy. Examples of parallel robots include the Delta and Steward Platform (see Fig. 3.3).

3.3.2 Disassembly Tools

The purpose of disassembly tools is to disestablish fasteners connecting the main components. The main components will be detachable after all corresponding connectors have been disestablished. The tools can be selected based on specific

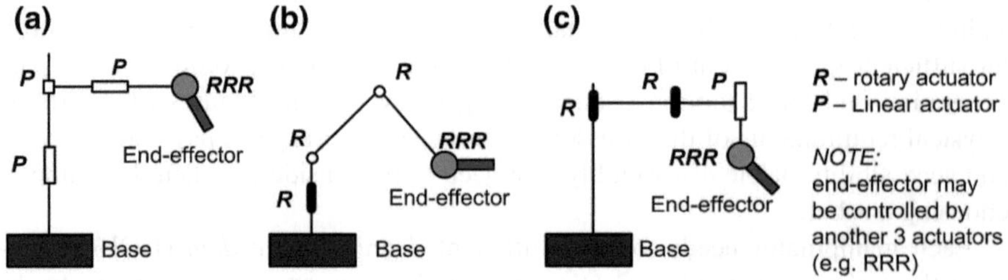

Fig. 3.2 Example of serial robots. **a** Cartesian robot. **b** Articulated robot. **c** SCARA robot

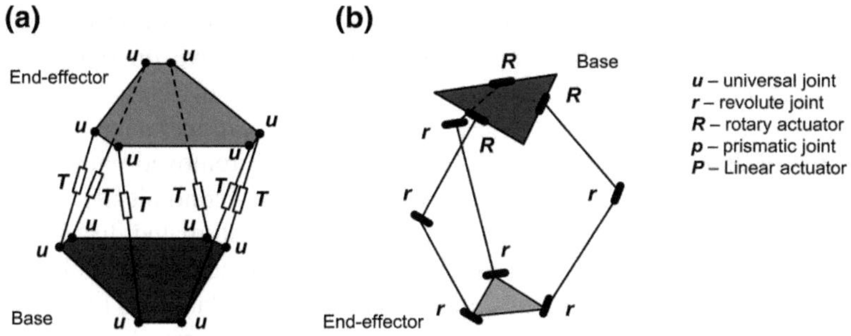

Fig. 3.3 Example of parallel robots. **a** Stewart platform. **b** Delta robot

tasks required for disestablish each type of fastener (see summary in Table 2.3). However, the required tool set differs between manual and automatic disassembly.

In manual disassembly, the human operator is flexible enough to utilise a limited set of primitive hand tools or power tools to disestablish a large variety of fasteners [8]. This is achievable thanks to our highly-developed senses and understanding of tool use: humans easily learn to control the tool appropriately in response to feedback, experiment to solve new problems using the fundamental knowledge of the tool's actions, and adapt their knowledge to new disassembly objects from past experience. Examples of such primitive tools include the screwdriver, pliers, cutters, grinder, saw and drill.

On the contrary, flexibility is quite limited in automated disassembly because of limitations in sensing and perception. Primitive disassembly tools may be able to use with disassembly automation by incorporating with proper sensors and control techniques. These can be complicated in most cases. Therefore, a number of attempts have been made to overcome this inflexibility with automatic tools that are specifically designed for each type of connectors. Lower-level control can be greatly simplified with tools that inherently required a straightforward method of application. Some innovative examples of such tools are as follows:

Automatic Screw Removal

Screws are a main category of fasteners and found in the majority of products. A number of special tools for the removal of screws have been developed. Screws can be removed by (a) *non-destructive*, (b) *semi-destructive,* and (c) *destructive* means.

For the *non-destructive approach*, an analysis of the unscrewing operation is done by Apley et al. [16]. The configuration of the prototype device is simple. A screwdriver is attached to an electric motor and a potentiometer is used for measuring the torque. The torque signal is continuously monitored to assess the condition of the unscrewing operation: whether the screw is coming out, the screwdriver is slipping on the head of the screw, or has completely missed the screw.

The condition is identified by using a least-square-based algorithm to analyse the measured torque signal. The success of the non-destructive operation relies on the accuracy in locating the screws, determining the type of screw head, and controlling the torque applied on it. These variations result in technical difficulties in automated disassembly. Even though the screws can be removed without damage, (semi-) destructive approaches are usually advantageous economically.

The *semi-destructive approach* is more feasible for automated disassembly due to its greater reliability. Seliger et al. [17] developed an automatic tool capable of unscrewing various types of screws, including those with damaged heads. The tool employs a pneumatic impact unit to create slots on the screw head. These slots become new active surfaces where torque can be transmitted, hence, allowing the unscrewing operation regardless of the shape and type of the screw. This tool is depicted in Fig. 3.4a, and the corresponding operation steps in Fig. 3.4b. Similarly, the "Drilldriver", developed by Feldmann et al. [18], utilises a similar principle of creating new active surfaces to overcome the problem of differing or damaged screw head shapes.

In the *destructive approach*, the screw head or surrounding component is intentionally damaged to disestablish the connection. This can be achieved using various methods of material removal, including with a chisel, milling, grinding, sawing or drilling.

Automatic Removal of Snap-Fits

Snap-fits are a low-cost option commonly used as fasteners according to the principle of Design for Assembly (DFA) [20]. Hook-like elements built into the main component elastically deform when pressure is applied in the assembly direction, and snap into place when the component is correctly positioned. In this manner, connections between main components can be established without using additional connective components and the assembly process is simplified. No special tool is required for assembly and the workspace may be smaller [7].

On the other hand, the difficulty arises in the disassembly process—especially for non-destructive disassembly. To non-destructively disassemble components connected by snap fasteners, multiple forces are required at different locations and in differing directions. Components that are designed to be disassembled may have a button which releases the corresponding snap-fit fittings. Where this is not the case, the tool end-effector must be small and the operation high in precision. Snap-fits are usually hidden and difficult to precisely locate. Manual detachment may be done by pulling, prying or deforming the components. However, these methods can easily damage components unless the worker is careful and skilled. Due to these difficulties, detaching components joined with snap-fits non-destructively is generally infeasible.

Automatic non-destructive disassembly currently only appears feasible in specific circumstances. Many have proposed the incorporation of new design elements in snap fits to facilitate disassembly (for further information, see *disassembly-embedded design* [21]). Fitting a magnet anchor on the flexible part of a snap-fit (see Fig. 3.5) allows the snap-fit to be pulled from outside. This principle can then be exploited using an automatic tool, as developed by Braunschweig [22].

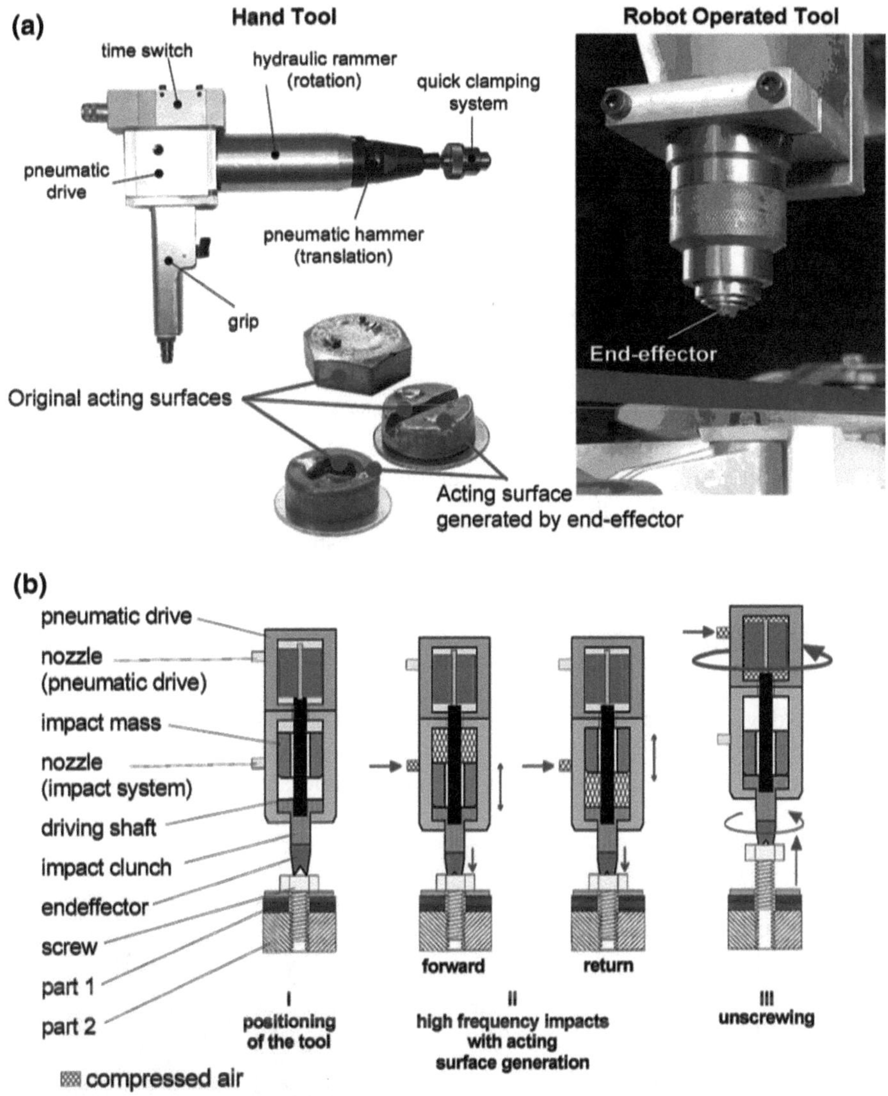

Fig. 3.4 Automatic unscrewer working by cutting a slot used as a new acting surface. **a** Two versions of tool developed from this concept: *a1* Hand tool—a hand held automatic unscrewer designed for manual use. *a2* Robot operated tool—an automatic unscrewer fully operated by a robot. **b** Operation steps for automatic unscrewing: positioning, creating acting surface, and unscrewing

Schumacher and Jouaneh [12] developed a prototype tool for disassembling snap-fit covers. The snap-fits in this case—a battery cover of electronic product—are not hidden and can be located by the vision system from the top view. The movement of this tool will be controlled by a robot. To detach snap-fits, the tool pushes the flexible part of the snap-fits until they are fully deflected. The exerted force is monitored by a force-sensing resistor (FSR) (see Fig. 3.6).

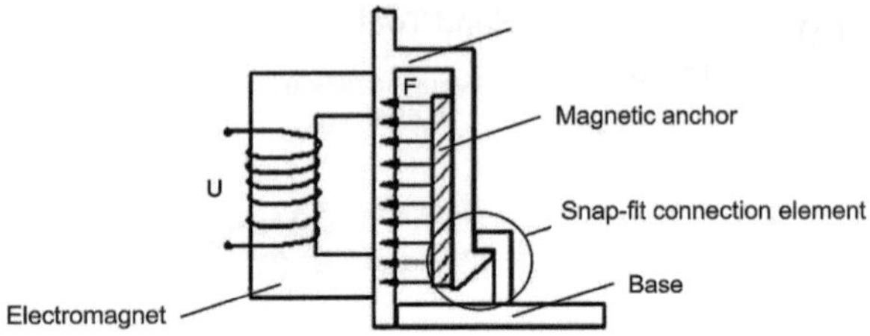

Fig. 3.5 Automatic snap-fits removal with electromagnetic trigger [22]

Other Semi-Destructive and Destructive Approaches

An aim of disassembly is to obtain some main components undamaged. (Semi-) destructive approaches, on the other hand, offer a reliable, robust and flexible method of separating materials joined by fasteners or attaining access to more valuable components that can be obtained non-destructively. Examples of automatic semi-destructive and destructive disassembly operations include using an angle grinder with an abrasive disc in the disassembly of LCD screens [15], and using plasma cutting to cut through the metal cover of washing machines [23].

3.3.3 Handling Devices

Handling devices refer to fixtures, grippers, and logistics systems used to control the motion of the parts and products according to the steps in disassembly process. The logistic system is used to convey the product into and out of the disassembly

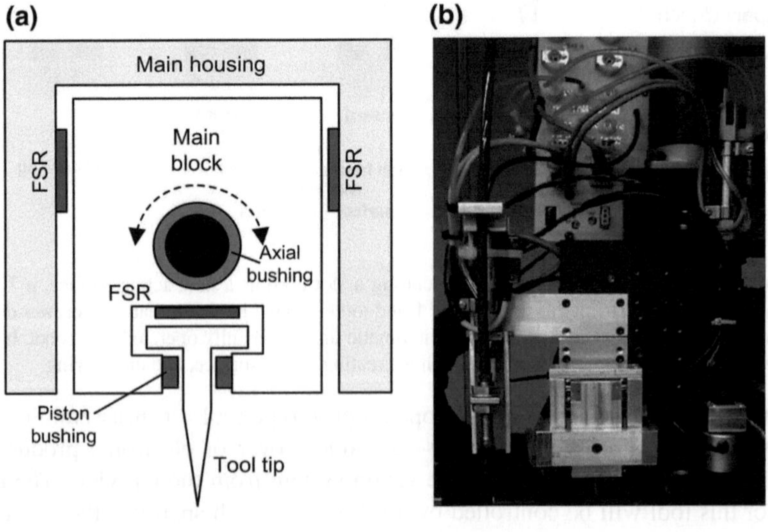

Fig. 3.6 Automatic snap-fits removal with FSR tool tip [12]. **a** Sketch of the disassembly tool. **b** Disassembly module

rig where the key activity of disassembly takes place. A variation of the designs of fixtures and grippers are used to overcome the uncertainties in the disassembly process. Meanwhile, the logistic system, e.g. conveyor belt and product storage, is straightforward due to its relatively simple function. Therefore, this section focuses on the fixture and gripper as opposed to the logistic system.

Product Fixtures

Fixtures are used to securely locate the product in specific locations and orientations during the disassembly process. Firm placement is essential in order to achieve high accuracy physical operations, and allow the necessary forces to be transmitted to the product. The product can be held with clamping techniques, e.g. using hydraulics or vacuum suction. If the product can be accurately positioned and is stationary with respect to the fixture, the exact position and orientation of the product can be exploited by the automatic system.

Alternatively, the fixture may be actuated to benefit the accessibility of tools in the disassembly operation. However, the difficulties are computation needed for coordination mapping and accumulating error from the structure of the moving fixture's components. A moving fixture is used in a number of systems, e.g. a rotating worktable that allows the robot to approach the back side of product (see Fig. 3.7a). A robot arm may also be used as the fixture [24] (Fig. 3.7b). In this case, there is a higher degree of freedom in the product location, however the product size and weight is limited by the specifications of the robot arm.

The fixture should be flexible enough to handle various product models, at least in the same product family. The components of the fixture should not obstruct the detection and removal of the product components and fasteners. One example of a product fixture designed for the disassembly application is the *flipping table* that is used in this research (see detail in Sect. 3.5).

Grippers

Various kinds of gripper are used in both assembly and disassembly processes to grasp and handle desired objects. A number of techniques by means of *hardware* and *software* are developed to overcome the uncertainties in geometry and operation.

Fig. 3.7 Fixture (adapted from [24, 25]). **a** Rotating worktable with a pneumatic clamp used for holding a product (computer case). The product can be rotated with 1-DOF. **b** Robot arm handling a product (wheel). The product is more accessible due to the 6-DOF robot's movement

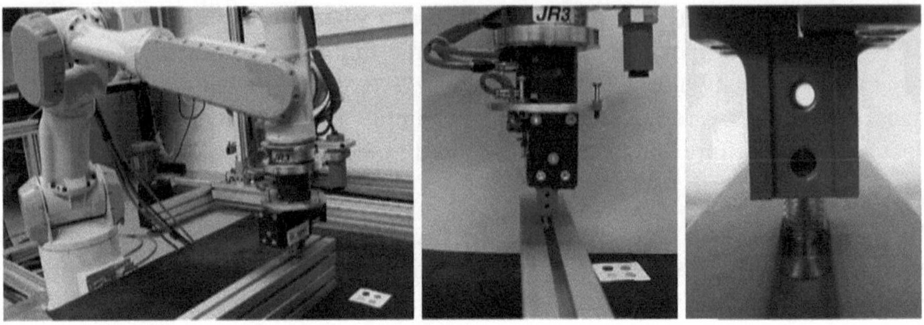

Fig. 3.8 Removal of a screw from a slot using force control [25]

In regard to the *hardware*, specific mechanisms are designed to suit requirements such as the specific geometry and features of the objects to be grasped. *Software*, in terms of control strategies, tends to relate to generic classes of grippers, e.g. 2-fingered-grippers. Inputs from vision and other sensor system are heavily used in this case. Focusing on the disassembly process, grippers are mainly used in two ways: (a) to disestablish connections and (b) to remove parts or components that have been detached.

For disestablishing connections, grippers are usually associated with action "pull" to detach the connectors from the remaining parts, for example pulling electrical/electronic cables [26]. A *fingered gripper* with force-torque control is needed because an accurate magnitude and direction of force and torque is crucial to avoid damage to the removed object. An example of this is in the removal of a screw from a rail slot in [25] (see Fig. 3.8).

For removing the detached parts or components, the gripper type is typically selected based on the geometry of the object. Much research has been done in developing strategies for handling objects with various shapes. The orientation of the gripper may be crucial in achieving a firm grasp and avoid collisions with other components. Visual input, together with force feedback, may be used to determine the size and shape of the object. Some work in this area is described as follows:

Schmitt et al. [5] developed the flexible gripper in Fig. 3.9 for the disassembly of Lithium-Ion batteries. The challenge here is to design a gripper to be flexible for grasping various geometries of battery cells. Two parallel jaw grippers are attached to a standard rail, providing the ability to adjust to the dimensions of the battery cell in two axes. In addition, voltage and resistance measurement at the battery contacts is done via the conductive contacts in the grippers' jaws.

Seliger et al. [27] developed a disassembly tool with a "screwnail" end-effector for removing the cover of products. The key advantage is to overcome the geometrical uncertainties of products. The operation procedure is shown in Fig. 3.10. The sharp "nail" tips secure the tool on the component rotationally, while the "screw" is driven into the material, securing it in the lateral direction. The screw is turned in the opposite direction to separate the tool from the product cover. This connection is strong enough to transmit the required forces and torques for disassembly.

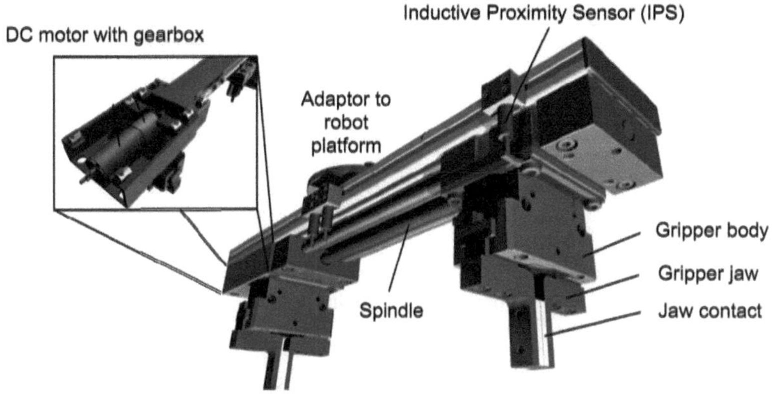

Fig. 3.9 Gripper for disassembly of Lithium-Ion battery [5]

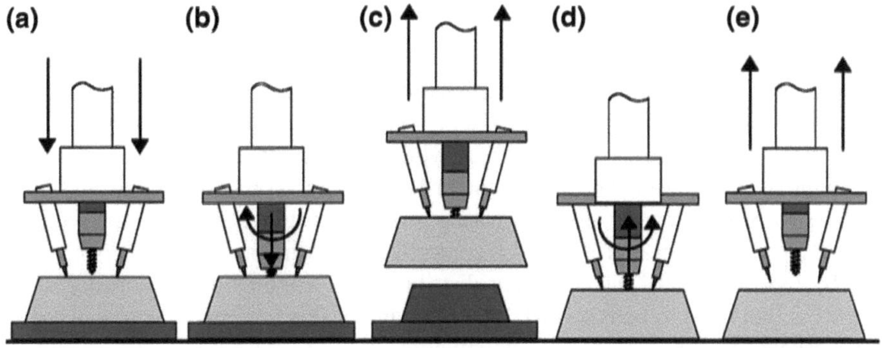

Fig. 3.10 Gripper for flexible disassembly using screwnail end-effector [27]. (**a–b**) Screwnail indention process. (**c–d**) Task execution process. **e** Separating process

In the area of software, Fernadez et al. [28] developed an algorithm to select appropriate grasping points for a multiple-fingered gripper to grasp objects of different geometries. Potential grips are assessed considering the centre of gravity of the object and friction cones calculated from the positioning of the fingers on the object (see Fig. 3.11). The system is learning-based and behaviour rules are extracted from the given training samples.

Disassembly automation also employs various types of grippers commonly used in object handling process. In addition to fingered grippers, *vacuum grippers* are commonly used in pick-and-place operations. Conventional robotic vacuum grippers consist of one or multiple air inlets connected to a source of negative pressure, which must be adequate to hold the object to the tool. Holding an object via one vacuum cup is susceptible to rotational moments, for example, caused by lack of alignment with the centre of gravity of the object. Using an array of cups or inlets eliminates this problem, however, requires additional assumptions regarding the shape and dimensions of the surface to be gripped.

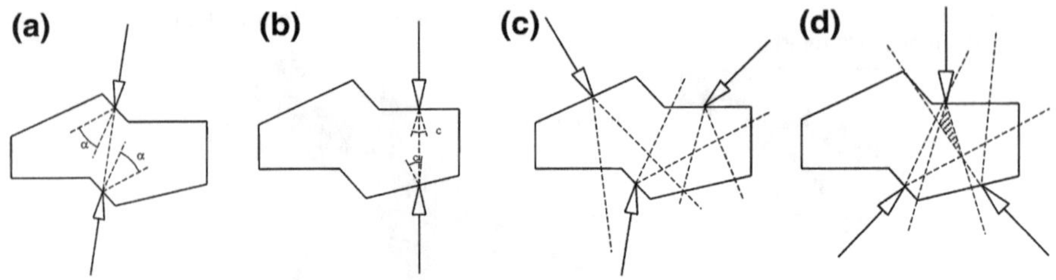

Fig. 3.11 Grasping point selection with two and three points contact [28]. **a** Two-point. **b** Two-point with force enclosure. **c** Three-point. **d** Three-point with force enclosure

3.4 Degree of Autonomy

3.4.1 Semi-automatic Disassembly

The *semi-automatic* or *hybrid disassembly system* consists of *automatic workstations* and *manual workstations* working in some kind of collaboration. The automatic workstation is equipped with sensors and operated by robot arms equipped with disassembly tools; meanwhile, human operators are employed at the manual workstation and carry out tasks that are more difficult to automate. A conveyor system can be used for transporting the products between the workstations. The main process flow can be controlled by an automatic planner or manually by the operators. The tasks are distributed to workstations according to the problems that need to be resolved. An advantage of this kind of system is in reducing workers' exposure to dangerous situations such as with heavy lifting and the removal of hazardous parts. On the other hand, their contribution allows the system to carry out processes that cannot be achieved with automation alone.

It is evident that this approach is, robust, cost-effective, and flexible for various product families [29]. Franke et al. [30] state that the hybrid disassembly process is economically justified; the manual operation is essential due to the inability of automated systems to operate with a non-determined disassembly sequence and react to varying product conditions. The, manual operation is, therefore, the fall-back option when the automatic operation has failed. Some research work with regard to hybrid disassembly systems is presented as follows.

Hybrid System for Disassembly of Various Product Families

At the *Institute for Machine Tools and Factory Management, Technical University of Berlin,* a hybrid system that is flexible to disassemble a wide range of product families has been developed. This work is presented in Kim et al. [29]. The study focuses on the automatic generation of plans and the control sequence of a system consisting of three robot arms and conveyor belts. The robots are responsible for heavy duty tasks, e.g. plasma cutting of the side wall of washing machines. Before the process starts, the system evaluates the degree of autonomy of the overall task from the product information and the availability of various systems.

Consequently, the tasks can be properly distributed to the manual and the automatic workstations.

Kim et al. [31] extended this concept to develop a disassembly line specific for LCD screens. At the manual workstation, the operator makes decision and disassembles monitor stands that interfere with other connectors that need to be removed. The LCD screen is carried to the automatic workstation by the robot. For the automatic workstation (see Fig. 3.12), the SCARA robot disassembles the components, i.e. screws, back cover, metal covers, printed circuit boards (PCBs), and cable connections, by using a two-fingered gripper. If the operations have failed, the product is forwarded to the next manual workstation to be further disassembled.

Hybrid System with Modular Concept

The *Institute for Handling Devices and Robotics (IHRT), Vienna University of Technology, Austria* has developed a multi-cell system designed for disassembling obsolete PCs. The work is presented in Kopacek and Kronreif [33]. The system consists of 2 robots and a manual workstation. Due to the problems in accessibility and various connection techniques, the operators manually detach the cables, the cover, and any valuable parts. The robots—equipped with a gripper and a screwdriver—remove other components in PC including the hazardous components. The planning is carried out by a semi-automatic planner, in which the specific operations and tool selections are made in advance.

The modular concept is applied to the semi-automatic system by Zebedin et al. [34] and Knoth et al. [6], which describes a system designed for extracting embedded components from PCBs. The operating modules (e.g. robot arms, part feeders, fixtures, desoldering system, and quality control system) are grouped as a subsystem. The disassembly cell controller is used to supervise the communication and co-ordination tasks in and between each subsystem. The cell controller is configured using hierarchical control and information distribution of the disassembly process. Machines can operate automatically and a human operator can command and monitor the system through the user interface.

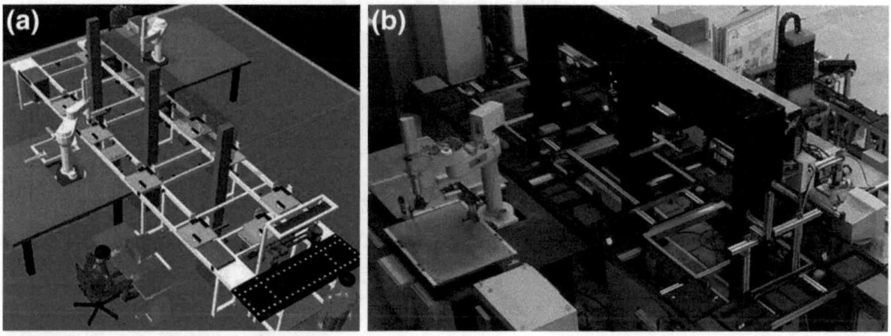

Fig. 3.12 Hybrid system for LCD screen disassembly [32]. **a** Verification in virtual system. **b** Automatic workstation in disassembly system

This system is also implemented as a part of the disassembly system for mobile phones described by Kopacek and Kopacek [9, 35], Kopacek [36]. The system employs 5 automatic workstations and a manual workstation. Automatic workstations carry out the tasks of cover removal, milling, drilling, and PCBs removal. Product feeding is done manually by the operator.

Disassembly Factories for Electrical and Electronic Products

Two disassembly systems at the *Institute for Machine Tools and Factory Management in Berlin, Germany* are presented in Basdere and Seliger [19].

First, a system for disassembling washing machines is presented in Uhlmann et al. [37]. The system consists of 2 stationary robots, 1 mobile robot, and 3 manual workstations (see Figs. 3.13 and 3.14). The automatic and the manual workstations are located in separated areas due to safety concerns. The products are transported between the workstations via a conveyor system. The robots perform plasma-cutting of metal covers and pass the product to the operators. Manual disassembly is done when the robots fail or the process is too complicated.

Second, a system for the disassembly of mobile phones is presented in Kniebel et al. [38]. The system consists of a 4-DOF SCARA robot and a manual workstation. The operator removes the battery then passes the product to the automatic workstation. The robot is equipped with a number of flexible tools, e.g. an automatic unscrewer, a vacuum gripper with a pin to open snap-fits, and a flexible clamp. The system is flexible enough to handle various models of mobile phones. A vision system is used to recognise the barcode in order to determine the model.

Other Selected Systems

Other interesting research is reviewed by Wicndahl et al. [39] and Scharke [7], which also illustrates the layouts of these systems.

Fig. 3.13 Hybrid system for disassembly of washing machines [19]

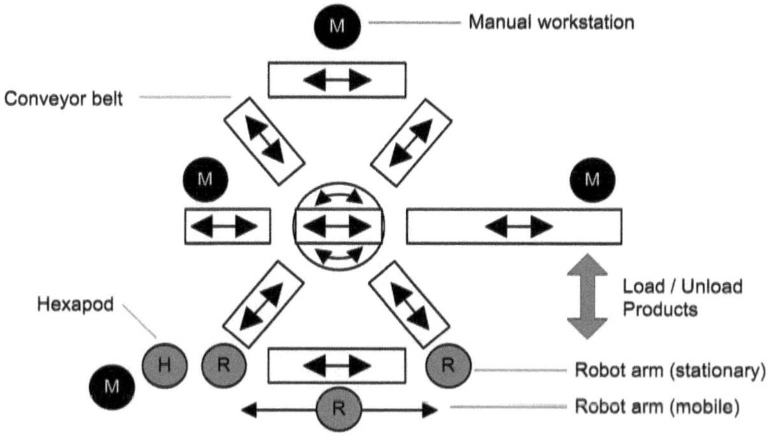

Fig. 3.14 Layout of the hybrid system for disassembly of washing machines (adapted from [39])

In the *Cleantech Project in Bochum, Germany*, a hybrid system for disassembling electrical plane and high pressure cleaning machines has been developed. This work is presented in Schnauber et al. [40]. The system consists of 2 manual workstations and an automatic workstation connected with a two-line structure conveyor system. The robot unfastens the screws at fixed locations. The first operator removes, separates, and sorts the parts according to reusability and material, while the second operator checks and cleans the parts. All disassembly plans are defined a priori.

At the *Fraunhofer Institute IML in Dortmund, Germany*, Jünemann et al. [41] presents a disassembly system consisting of 2 manual workstations and an automatic workstation. The products are transferred between the workstations with a closed-loop conveyor system. The robot is used to remove the cover of products, e.g. microwave ovens. The operators manually separate the valuable parts from the product.

At the *Institute for Production Automation and Handling Technology at the TU-Braunschweig, Germany*, a system for separating valuable and hazardous components from PCBs was developed. This work is presented in Hesselbach et al. [42], Friedrich [43]. The system consists of an automatic workstation and a manual workstation. The operator detaches cables and connectors, and then passes the PCB to the automatic workstation. At the automatic workstation, a SCARA robot separates the components using a special gripper for microchips and an unsoldering tool. A laser scanner identifies the embedded components on PCBs based on a component database. Subsequently, the robot tasks are generated by a planning and control system.

At the *Technion's Center for Manufacturing Systems and Robotics, Haifa, Israel,* a system has been developed consisting of 2 automatic workstations and a manual workstation. The automatic workstations are operated by a 6-DOF robot and a 4-DOF robot. The manual workstation is used to deal with complicated tasks.

The system is designed to work with multiple pallets at the same time where the pallet can be identified using electronic non-contact badge. The related research in vision systems and advanced planning is also described in Zussman [44].

3.4.2 Fully Automatic Disassembly System

The heightened degree of autonomy in *fully automatic disassembly systems* is generally achieved by incorporating (a) better sensor modules, (b) prior knowledge of products, and (c) a high-level task planner. Ideally, the intelligent planner is able to control the system to carry out the operations and address all uncertainties during the process, therefore eliminating the need for human intervention.

However, due to the variations in EOL products and the uncertainties in the disassembly process, achievement of the required flexibility and robustness has posed to be a great challenge. As one cannot guarantee that all problems can be resolved solely by automation, it is expected that minimal human intervention may be involved. However, unlike semi-automatic disassembly, operators in a fully automatic disassembly system do not carry out tasks on the product directly but monitor and maintain the system and provide instructions at the planning level. This can be considered a type of human-machine collaboration. The relevant research work is presented as follows.

Co-operative Multi-sensorial Disassembly Cell

The *Department of Physics, Systems Engineering and Signal Theory*, *University of Alicante*, *Spain*, Torres et al. [45] has developed one of the most advanced disassembly cells for disassembling computers to date (see Fig. 3.15). This disassembly cell consists of two industrial articulated robots equipped with force-torque sensors and selected interchangeable disassembly tools. This project comprises three major contributions: (a) co-operative operation, (b) vision system and (c) multi-sensorial system.

Both robots work co-operatively through a task planner that automatically generating paths and trajectories based on a graph model proposed by Torres et al. [46]. In terms of the vision system, work has been described in the detection of occlusions [47] and various components including screws and cables [25]. For the multi-sensorial system, Gil et al. [25] combines information between a tactile sensor and the vision system in order to perform visual servoing of the robot. This system was tested by removing a bolt from a straight slot (see Fig. 3.8).

This system solves the problem of uncertainties at the operational level by using an integrated sensor system. However, at the higher level, all disassembly sequence planning is based on precedence among assemblies [48]. Hence, the user is still required to provide this specific product structure information a priori. Input from the vision system is only used to determine the detailed geometry for the operation level.

Fig. 3.15 Multi-sensorial cooperative robots—demonstrate removing CD drive [25]

Automatic System for the Disassembly of EOL Personal Computers

At *the University of Bridgeport, USA*—ElSayed et al. [13], as system has been developed for the selective disassembly of EOL personal computers (PC) for reuse and recycling purposes. The system consists of an articulated industrial robot and a camera system, and deals with uncertainties in the product structure using (a) a vision system and (b) an *online disassembly sequence planner using genetic algorithms* (GA).

The vision system consists of a 2D-camera and a laser range sensor. Camera images are used to recognise and locate the components, matching regions of the input image with predefined 2D templates of the components supplied in the bill-of-materials (BOM). The BOM of electrical/electronic products are subject to change during the usage stage due to modifications. PCs are one example where the location and the existence of the main components (e.g. random access memory (RAM)) and connectors (e.g. screws) are likely to change according to upgrading, repairing, or personal preference. This creates uncertainties in the product structure at the point of disassembly.

These uncertainties are addressed by the online disassembly sequence planner). The supplied BOM contains all expected components and the precedence relations between them. At any stage, and according to any modifications to the product, only a select number of these components are detected by the system. From the detected components, the GA implementation generates a (near-)optimal solution for the disassembly sequence, minimising a certain criteria, e.g. travel distance.

The key feature of this system is its ability to adapt the plan according to the current situation perceived by the vision system. However, a BOM representing the precise product structure and a template of specific components needs to be supplied a priori.

Automatic System for the Disassembly of Automotives

At *Heinz Nixdorf Institute, University of Paderborn, Germany*, a disassembly system for wrecked cars has been developed. The work is presented in Büker et al. [24]. This research is part of Project DEMON founded by the German Ministry for Education and Research (523-4001-01 IN 506 B 2). This research focuses on the disassembly of automotive wheels with variation in the size of the wheels, the number of the bolts, and the position of the wheel. The system employs active stereo cameras and a robot arm equipped with a tool for unscrewing (see Fig. 3.7b).

The active stereo cameras are used to reconstruct the 3D structure of the product. Principle component analysis (PCA) is used to identify components that are difficult to recognise due to uncertainties in the EOL condition, e.g. due to rust. A knowledge-based approach with a neural network is used to address the problem of occlusion in complex scenes [49].

Other Selected Systems

At the *Automation and Control Institute, Vienna University of Technology, Austria*, Merdan et al. [50] proposes an ontology-based architecture with a multi-agent system (MAS) for the disassembly of digital cameras. The robot disassembles the product using a screwdriver and grippers. The vision system detects the components and links to the knowledge base. The ontology is used to describe the tasks and their requirements according to the operating modules and level of control. The disassembly plan is automatically generated from the hierarchical product structure described in the ontology. Subsequently, the operation plan with optimised tool-paths can be generated.

At the *Department of Manufacturing and Mechanical Engineering, Miami University, USA*, Bailey-Van Kuren [51] proposed a strategy for the real-time generation of cutting tool paths which need to follow the complex surface of a product. A prototype system consists of a robot equipped with a cutting tool, a vacuum gripper, and the smart vision system based on structured light [52]. The mobile phone is the case-study product in this research.

At the *Fraunhofer Application Centre for Logistic System Planning and Information System, University of Technology, Cottbus, Germany*, Scholz-Reiter et al. [14] conducted research in collaboration with *Prielog Logistik GmbH* [53] where the disassembly system is developed. The system designed for the disassembly of obsolete TVs and monitors into their components within 5–7 min. The system is operated by two robots. The first one is a 6-DOF robot equipped with various disassembly tools for destructive and non-destructive disassembly. The second is a 4-DOF robot equipped with different grippers for handling parts. Vision is used to recognise variations in the product and component condition. Flexible disassembly plans are generated according to the sensed information and the existing database. The method to generate reactive and adaptive disassembly plans from the data is described in Scharke [7].

At the *Fraunhofer Institute for Information and Data Processing (IITB), Karlsruhe, Germany*, a system for disassembling used cars is developed based on MAS. The work is presented in Gengenbach et al. [54]. The system consists of 2 robots: one for disassembling and the other for monitoring. The disassembly robot is equipped with an electric screwdriver and gripper. The "eye-in-hand" approach is implemented on the monitoring robot, which manipulates a miniature stereo camera system, called MiniVISTAs. Tonko and Nagel [55] focused on the vision system, developing an algorithm for the model-based tracking of a non-polyhedral car engine.

At the *Fraunhofer-Institute for Production and Automation in Stuttgart, Germany*, two systems were developed. First, a system for disassembling telephones is presented in Kahmeyer [56]. The robots are equipped with various tools: a flexible parallel gripper, a vacuum gripper, an unscrewer, a tool for detaching snap-fits, and a drill. The system is able to carry out low-level planning and control. It can be adapted to different phones with manual modification. Second, a system for disassembling roofs and windows of cars is presented in Rupprecht [57]. A vision system is used to locate the position of these components, which are removed by two robots equipped with special disassembly tools.

At the *Chair of Manufacturing Automation and Production at University Erlangan-Nürnberg, Germany*, a system for separating the embedded components on PCBs by flexible interconnected disassembly cells has been developed. The work is presented in Feldmann and Scheller [58]. The system uses a laser scanner to identify the embedded components. The valuable components are disassembled non-destructively while the hazardous components are disassembled using cutting techniques.

3.5 System Setup in This Research

In this research, at the *School of Mechanical and Manufacturing Engineering, The University of New South Wales, Australia*, Vongbunyong et al. [15] has developed a low-cost automated disassembly system in which LCD screens are used as the test case. A (semi-) destructive approach is implemented to selectively disassemble to a module level. The principle of cognitive robotics is used to resolve the uncertainties in product and process. As a result, the system is flexible to disassemble various models of products regardless of the exact information in product structure and geometry. In addition, the system is capable of learning from past experience, which results in an increase of performance of the process.

The system is controlled by the *cognitive robotic agent* (CRA) which is an intelligent planner with cognitive functions. The CRA is part of the cognitive robotic module (CRM). The CRM also consists of a knowledge base (KB), where learned information is stored. A vision system module (VSM) is designed for detecting the components and assesses the success or failure of an operation. The VSM acquires and processes images from a colour camera and a depth sensor. The mechanical components and their low-level control will in this case be called the disassembly operation unit modules (DOM). The system employs a 6-DOF robot equipped with an angle grinder to perform (semi-)destructive disassembly. To reduce the complexity of using a tool changer, the *flipping table* is designed to function as the fixture and the means of removing a detached component from the product.

For the overview, the architecture of the system according to the operating modules and level of control is shown in Fig. 3.16. The communication system and information flow are further explained in Sect. 6.4.1 (Fig. 3.17).

Robot Arm

The robot arm is a stand-alone module used as the primary component of the DOM. This module consists of 2 elements: (a) *IRB-140*, a small-sized 6-DOF articulated industrial robot and (b) *IRC-5* robot controller. The system can be programmed using the high-level procedural language RAPID. The main task is to control the disassembly tool to perform the (semi-) destructive disassembly operation according to the primitive action requested by the CRA. The robot has built-in sensors used for preventing an overload on the motor at each joint. *MotionSupervision* [59] is implemented using these sensors and allow the system to recognise collisions. This collision awareness is used in finding a set of process parameters that allow the disassembly tool to approach the required cutting destination.

Flipping Table and Grinder

The *flipping table* is a device used for handling the sample to be disassembled and removing the detached components. This device is a low-cost substitution to classical grippers, designed to remove the detached parts from the product without

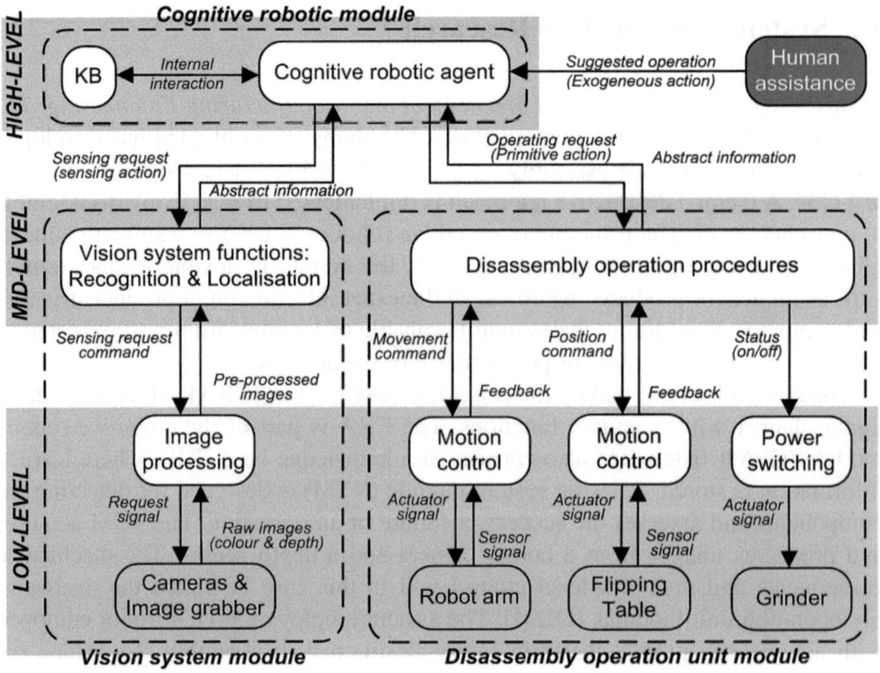

Fig. 3.16 Example of system architecture—levels of control and operating modules

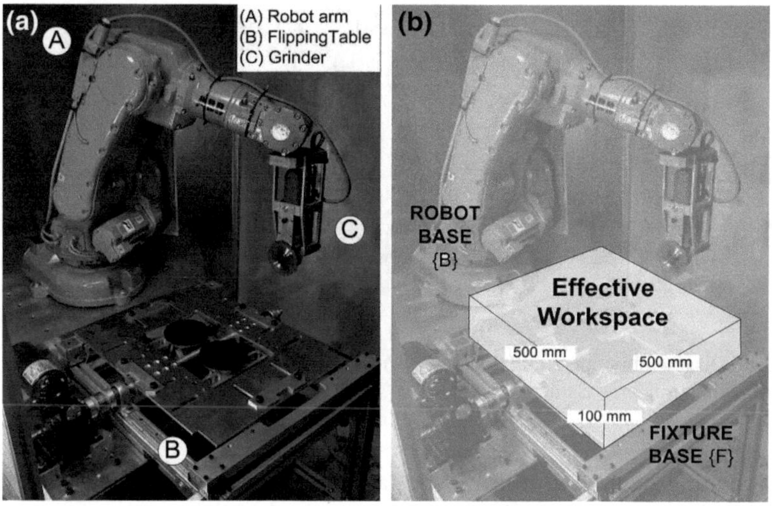

Fig. 3.17 Disassembly rig for disassembly of LCD screens. **a** Complete setup excluding cameras. **b** Effective workspace

gripping. This solution addresses the complexity of picking up objects of varying geometry amidst clutter; however, the ability of the gripper to selectively pick an object or to apply forces in specific directions is lost.

An LCD screen sample is initially placed firmly on the fixture plate and gripped from the front side using two suction cups. The suction cup generates the axial force while the fixture elements situated on each side prevent any rotation or transverse movement (see Fig. 3.18a). The fixture plate is designed to support 16″–19″ LCD screen. By gripping the sample from the front side, it is ensured that the LCD module is the last component remaining on the workstation after disassembly. The flipping table is activated after each operation cycle by flipping the entire product by almost 180°. As a result of the flipping force and gravity, any components and detached parts are unloaded into a container or conveyor situated underneath.

Disassembly Tool

The angle grinder is used as the disassembly tool for the cutting operation. (Semi-) destructive disassembly has been implemented due to its high success rate and economic feasibility. A grinder equipped with a multi-purpose abrasive cut-off disc is a versatile cutting tool capable of cutting all materials found in LCD screens (e.g. steel, aluminium, polymers; see Fig. 3.18b) without losing its sharpness. However, a major drawback is tool wear. As a result, the tool-tip position changes non-uniformly according to the material, cutting speed, and feed speed. The vision system is used to sense the current tool-tip position in order to compensate for the change.

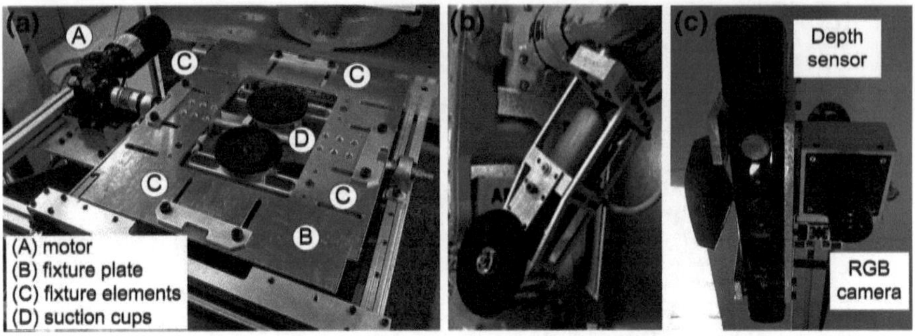

Fig. 3.18 Disassembly operation module (**a, b**) and vision system module (**c**). **a** Flipping table. **b** Robot with a grinder. **c** Camera system

The cutting operation destroys the fastener connections without any need to adapt the physical tooling, allowing the system to handle screws, wires as well as hidden snap fit connections by cutting through main components at less critical locations near the border. For this case-study product, a vertical cutting direction is sufficient for the disassembly process.

In conclusion, instead of grippers and multiple disassembly tools, a small set of devices has been selected in order to reduce cost and computing resources. The added complexity of force-torque control and tool-change has been eliminated while still allowing the system to be robust and flexible to physical uncertainties. However, a major drawback is component damage from the disassembly operation; this solution is suitable for recycling purposes only.

3.6 Conclusions

Disassembly automation generally refers to an automated system consisting of robot arms, equipped with disassembly tools, handling devices, and sensors. Automation is implemented in the disassembly process to address the problem in economic feasibility directly related to high labour costs, especially in developed countries. Disassembly automation can also be used to aid or replace the operator in performing undesirable tasks, for instance handling heavy parts or those that contain hazardous substances. Human operators are still required to various extents to aid the system in planning and/or carry out activities that the automation lacks the flexibility for.

In *semi-automatic* or *hybrid disassembly*, human operators at manual workstations work in collaboration with the automatic workstation in executing physical tasks on the product or component. The major advantage is the high flexibility in dealing with uncertainties and variations in the products. Economic feasibility may be achieved with the automatic workstation performing certain tasks more cheaply and efficiently, while the human operators provide the flexibility to recover from problems where the automation fails. However, in this situation the operators are still directly exposed to the product, which in some cases may be harmful.

Automatic disassembly systems carry out the process more autonomously, and are aimed at executing disassembly processes with minimal human involvement. Greater autonomy at various levels can be achieved by employing more intelligent planners and smarter sensor systems; this remains an active research area. The role of human operators becomes supervisory; exposure to potentially-harmful conditions can be reduced. Problems due to the variations in EOL products and uncertainties in the disassembly process still remain not completely addressed. Most existing research has focused on low-level operations, e.g. tool path generation. At the planning level, it is clear that prior knowledge of the product structure always needs to be supplied in various forms, e.g. CAD models, a BOM, and/or precedence relations. The high-level planner needs this specific information for each product model to generate the DSP and DPP. The flexibility of the system to handle variations in models and products is limited as this information is generally unavailable before a product has been disassembled.

In conclusion, it is desirable, both in the interest of preserving the safety of human workers and of achieving economic feasibility, to continue the path of realising fully automatic disassembly systems. This research aims to increase the flexibility of fully automatic systems to the variety of product models, and their robustness, to the uncertainties in the process. Plans need to be generated by the intelligent agent regardless of uncertainties in the specific model information. In this book, the principle of cognitive robotics is implemented to resolve these uncertainties (see Chaps. 5 and 6).

References

1. Rifkin J (1995) The end of work: the decline of the global labor force and the dawn of the post-market era. Putnam Publishing Group, New York
2. Bennett S (1993) A history of control engineering 1930–1955. Peter Peregrinus Ltd., London
3. Craig JJ (2005) Introduction to robotics: mechanics and control, 3rd edn. Prentice Hall, Englewood Cliffs
4. Duflou JR, Seliger G, Kara S, Umeda Y, Ometto A, Willems B (2008) Efficiency and feasibility of product disassembly: a case-based study. CIRP Ann Manuf Technol 57(2):583–600
5. Schmitt J, Haupt H, Kurrat M, Raatz A (2011) Disassembly automation for lithium-ion battery systems using a flexible gripper. Paper presented at the IEEE 15th international conference on advanced robotics: new boundaries for robotics (ICAR 2011)
6. Knoth R, Brandstötter M, Kopacek B, Kopacek P (2002) Automated disassembly of electr(on)ic equipment. In: IEEE international symposium on electronics and the environment
7. Scharke H (2003) Comprehensive information chain for automated disassembly of obsolete technical appliances. GITO-Verlag, Berlin
8. Lambert AJD, Gupta M (2005) Disassembly modeling for assembly, maintenance, reuse, and recycling. CRC Press, Boca Raton

9. Kopacek P, Kopacek B (2006) Intelligent, flexible disassembly. Int J Adv Manuf Technol 30(5–6):554–560
10. Kopacek B, Kopacek P (2001) Semi-automatised disassembly. In: The 10th international workshop on robotics, Alpe Adria Danube region, RAAD 01, Vienna, pp 363–370
11. Golnaraghi F, Kuo BC (2009) Automatic control systems, 9th edn. Wiley, London
12. Schumacher P, Jouaneh M (2013) A system for automated disassembly of snap-fit covers. Int J Adv Manuf Technol 1(15):2055–2069
13. ElSayed A, Kongar E, Gupta SM, Sobh T (2012) A robotic-driven disassembly sequence generator for end-of-life electronic products. J Intell Robot Syst Theor Appl 68(1):43–52
14. Scholz-Reiter B, Scharke H, Hucht A (1999) Flexible robot-based disassembly cell for obsolete TV-sets and monitors. Robot Comput Integr Manuf 15:247–255
15. Vongbunyong S, Kara S, Pagnucco M (2013) Application of cognitive robotics in disassembly of products. CIRP Ann Manuf Technol 62(1):31–34
16. Apley DW, Seliger G, Voit L, Shi J (1998) Diagnostics in disassembly unscrewing operations. Int J Flex Manuf Syst 10(2):111–128
17. Seliger G, Keil T, Rebafka U, Stenzel A (2001) Flexible disassembly tools. In: IEEE international symposium on electronics and the environment, pp 30–35
18. Feldmann K, Trautner S, Meedt O (1996) Innovative disassembly strategies based on flexible partial destructive tools. Annu Rev Control 23:159–164
19. Basdere B, Seliger G (2003) Disassembly factories for electrical and electronic products to recover resources in product and material cycles. Environ Sci Technol 37(23):5354–5362
20. Boothroyd G, Alting L (1992) Design for assembly and disassembly. CIRP Ann Manuf Technol 41(2):625–636
21. Masui K, Mizuhara K, Ishii K, Rose C (1999) Development of products embedded disassembly process based on end-of-life strategies. In: The EcoDesign'99: 1st international symposium on environmentally conscious design and inverse manufacturing, Tokyo, pp 570–575
22. Braunschweig A (2004) Automatic disassembly of snap-in joints in electro-mechanical devices. In: The 4th international congress mechanical engineering technologies'04, Varna
23. Uhlmann E, Spur G, Elbing F (2001) Development of flexible automatic disassembly processes and cleaning technologies for the recycling of consumer goods. In: IEEE international symposium on assembly and task planning, pp 442–446
24. Büker U, Drüe S, Götze N, Hartmann G, Kalkreuter B, Stemmer R, Trapp R (2001) Vision-based control of an autonomous disassembly station. Robot Auton Syst 35(3–4):179–189
25. Gil P, Pomares J, Puente SVT, Diaz C, Candelas F, Torres F (2007) Flexible multi-sensorial system for automatic disassembly using cooperative robots. Int J Comput Integr Manuf 20(8):757–772
26. Kernbaum S, Franke D, Seliger G (2009) Flat screen monitor disassembly and testing for remanufacturing. Int J Sustain Manuf 1(3):347–360
27. Seliger G, Stenzel A, Rebafka U, Zuo B-R (2002) A novel disassembly tool with screwnail end effectors. J Intell Manuf 13(3):157–163
28. Fernandez C, Reinoso O, Vicente MA, Aracil R (2006) Part grasping for automated disassembly. Int J Adv Manuf Technol 30(5–6):540–553
29. Kim H-J, Harms R, Seliger G (2007) Automatic control sequence generation for a hybrid disassembly system. IEEE Trans Autom Sci Eng 4(2):194–205
30. Franke C, Kernbaum S, Seliger G (2006) Remanufacturing of flat screen monitors. In: Brissaud DTS, Zwolinski P (eds) Innovation in life cycle engineering and sustainable development, pp 139–152
31. Kim H-J, Chiotellis S, Seliger G (2009) Dynamic process planning control of hybrid disassembly systems. Int J Adv Manuf Technol 40(9–10):1016–1023
32. Kim HJ, Kernbaum S, Seliger G (2009) Emulation-based control of a disassembly system for LCD monitors. Int J Adv Manuf Technol 40(3–4):383–392
33. Kopacek P, Kronreif G (1996) Semi-automated robotized disassembling of personal computers. Paper presented at the IEEE symposium on emerging technologies and factory automation (ETFA 2)

34. Zebedin H, Daichendt K, Kopacek P (2001) A new strategy for a flexible semi-automatic dis-assembling cell of printed circuit boards. IEEE Int Symp Ind Electron 3:1742–1746
35. Kopacek P, Kopacek B (2003) Robotized disassembly of mobile phones. Paper presented at the IFAC workshop: intelligent assembly and disassembly (IAD'03), Bucharest
36. Kopacek P (2005) Semiautomatized disassembly—some examples. In: IFAC proceedings volumes (IFAC-PapersOnline) 16:146–151
37. Uhlmann E, Haertwig JP, Seliger G, Keil T (2000) A pilot system for the disassembly of home appliances using new tools and concepts. In: Proceedings of the third world congress on intelligent manufacturing processes and systems, Cambridge, MA, 28–30 June 2000
38. Kniebel M, Basdere B, Seliger G (2004) Hybrid disassembly system for cellular telephone end-of-life treatment. In: The joint international congress and exhibition electronics goes green, Stuttgart, pp 281–286
39. Wicndahl H-P, Scholz-Reiter B, Bürkner S, Scharke H (2001) Flexible disassembly systems-layouts and modules for processing obsolete products. Proc Inst Mech Eng Part B J Eng Manuf 215(5):723–732
40. Schnauber H, Kiesgen G, Slawik F (1997) Das element produkt-wiederverwendung im qualitätskreislauf. Ergebnisse des EUREKA Forschungsprojektes CLEANTECH (EU 1104)—concept for logistical and environmental disassembly technologies. Forschungszentrum Karlsruhe GmbH, Karlsruhe
41. Jünemann R, Hauser H, Moukabary G (1997) Sensor- und informationssystemein der demontage. Paper presented at the in colloquium on closed-cycle economy and disassembly, Berlin, Germany, 30–31 Jan 1997
42. Hesselbach J, Friedrich R, Schütte G (1994) Automamization in dismantling of printed circuit boards. Paper presented at the international seminar on life cycle engineering, RECY'94, Erlangen, Germany
43. Friedrich R (1996) Identifizierung elektronischer bauelemente und dren gezielte demontage. VDI-Verlag
44. Zussman E (1995) Planning of disassembly systems. Assembly Autom 15(4):19–22
45. Torres F, Gil P, Puente ST, Pomares J, Aracil R (2004) Automatic PC disassembly for component recovery. Int J Adv Manuf Technol 23(1–2):39–46
46. Torres F, Puente S, Díaz C (2009) Automatic cooperative disassembly robotic system: task planner to distribute tasks among robots. Control Eng Pract 17(1):112–121
47. Gil P, Torres F, Ortiz FG, Reinoso O (2006) Detection of partial occlusions of assembled components to simplify the disassembly tasks. Int J Adv Manuf Technol 30(5–6):530–539
48. Torres F, Puente ST, Aracil R (2003) Disassembly planning based on precedence relations among assemblies. Int J Adv Manuf Technol 21(5):317–327
49. Büker U, Hartmann G (1996) Knowledge based view control of a neural 3-D object recognition system. Paper presented at the 13th international conference on pattern recognition
50. Merdan M, Lepuschitz W, Meurer T, Vincze M (2010) Towards ontology-based automated disassembly systems. In: Industrial electronics conference (IECON), pp 1392–1397
51. Bailey-Van Kuren M (2006) Flexible robotic demanufacturing using real time tool path generation. Robot Comput Integr Manuf 22(1):17–24
52. Bailey-Van Kuren M (2005) A demanufacturing projector–vision system for combined manual and automated processing of used electronics. Comput Ind 56(8–9):894–904
53. Hucht A (1996) Automatisierung: Roboter zerlegen Elektronikschrott. Umwelttechnik 2:36
54. Gengenbach V, Nagel H-H, Tonko M, Schaefer K (1996) Automatic dismantling integrating optical flow into a machine vision-controlled robot system. In: IEEE international conference on robotics and automation, pp 1320–1325
55. Tonko M, Nagel H-H (2000) Model-based stereo-tracking of non-polyhedral objects for automatic disassembly experiments. Int J Comput Vis 37(1):99–118
56. Kahmeyer M (1995) Flexible Demontage mit dem Industrieroboter am Beispiel von Fernsprech-Endgeräten. IPA-IAO Forschung und Praxis. Springer, Germany
57. Rupprecht R (1998) Flexibel automatisierte demontage von fahrzeugdachern. IPA-IAO—Forschung und Praxis. Springer, Berlin

58. Feldmann K, Scheller H (1994) Partial automated disassembling of used electronic products and their components. Paper presented at the IEEE international symposium on electronics and the environment, San Francisco, USA
59. ABB (2004) RAPID overview the heart of robotics—RAPID reference manual, 3HAC16580-1 Revision C edn.

Chapter 4
Vision System

Abstract Any system that needs to make dynamic decisions based on a variable situation requires some form of perception. The system must first determine what the current situation is before it can appropriately react to this situation. The system of sensing, recognising, and localising objects of disassembly, its components and/or fastening elements at a distance is referred to here as the *vision system*. This chapter provides an introduction to the requirements and considerations for the implementation of a vision system for autonomous disassembly.

4.1 Introduction

4.1.1 Why a Vision System?

Experience in the field of robotic disassembly has indicated that robotic disassembly cannot simply be considered as the reversal of assembly [1], due to two main factors: the use of irreversible fasteners and the presence of a higher degree of uncertainties. While the former can only be handled with (semi-) destructive disassembly and an appropriate choice of hardware, appropriate compensation for the latter falls largely within the domain of the vision system.

An accurate world model is required in order to respond appropriately to the given situation. If the existence, location, condition and tolerances of all components can be guaranteed, as may be the case in *assembly*, it may be sufficient to employ an autonomous system that is "blind" and relies entirely on internal knowledge. For *assembly*, because lot sizes are known in advance, it may be feasible to reduce job complexity and increase accuracy by designing appropriate jigs and fixtures specific to the product design. A high emphasis on precision leads to an almost deterministic execution of programs. Small deviations may be compensated using force and torque sensors; however, one expects that all fundamental knowledge, including the effect of actions is correct.

S. Vongbunyong and W.H. Chen, *Disassembly Automation*, Sustainable Production, Life Cycle Engineering and Management, DOI 10.1007/978-3-319-15183-0_4

It is often uneconomical to design a *disassembly* system to suit a particular product design. Flexible methods of fixing the product in the workspace may mean that errors in product location are higher. A variety of products are expected, in unpredictable lot sizes, arriving with varying end-of-life product conditions. A commonly-successful disassembly operation may fail due to rust or damage during the usage phase or subsequent transport. External appearance is often insufficient to predict such results ahead of time. Due to repairs or modifications during the usage phase, components may also be exchanged, added or missing. These problems can be broadly summarised by saying that there is an increased degree of *uncertainties* in disassembly. A smarter system is required, capable of reacting to deviations in the object of disassembly, rather than relying entirely on static knowledge [2].

In this book, the broad term *vision* is used to describe all scene sensing techniques, including those that provide or use distance data instead of light intensity. The employment of a vision system enables the system to deal with such uncertainties by updating the agent with knowledge specific to an object to be disassembled. A vision system consists of:

- A method of obtaining *actual geometric data* of a scene from a distance
- The processing of raw sensory data to *semantic information* regarding the contents of the scene, e.g. detected components and their location.

The automatic localisation of products and components enables the system to be more flexible to positioning errors. Automatic identification of products reduces the effort required in setting the system up for each product's arrival. Automatic detection of components allows the system to dynamically respond to product modifications. The success or failure of operations can be verified by *seeing* whether the goal state has been achieved, and in the case of failure, back-up measures can be taken. Furthermore, the ability to detect relevant disassembly elements opens the possibility for the disassembly of previously-unknown product variants, using only knowledge of the generic structure of a product type.

However, vision processing is a non-trivial task and a continually-developing research area. Potential solutions are often limited by high processing times and/or inappropriately high failure rates. The following sections provide an overview of the considerations involved in the implementation of a vision system for disassembly robotics.

Foreseeable alternatives for a "blind" disassembly systemalso exist, with a lower degree of autonomy, lower speed and higher tooling and labour requirements. For completeness, one such alternative is as follows:

- A generic product is manually positioned; the hardware setup ensures that the product remains in a known position for the entire procedure.
- A human worker inputs the product ID, and if necessary, the current stage of disassembly.
- The robot is pre-programmed with "skills" such as gripping and unscrewing, which are defined entirely by an initial tool position and subsequent contact sensing.

- To program or "teach" new action instances, the robot is manually positioned for the first execution of the skill. Parameters (e.g. location of contact surfaces) are saved for future execution.
- If the robot has already been programmed for this disassembly stage, the programmed actions are automatically executed.
- When unexpected forces are detected, the program terminates automatically with a call for assistance. Touching the "wrong" surface with the tool must be inconsequential.

However, such a solution requires for the entire procedure to be manually taught with the introduction of each new product variant, and is not robust to product modifications, inducing additional labour costs. Every incoming product must be manually identified. Manual positioning may be both slow and physically demanding. A lack of vision constrains the robotic system from further autonomy.

4.1.2 General Requirements

The function of the vision system is to obtain relevant scene information from the real environment, so that the system is robust to the aforementioned uncertainties. This consists primarily of the *detection* of disassembly-relevant objects, which entails

- *Recognition*: identifying or classifying the respective object; and
- *Localisation*: obtaining the coordinates of the object in 3D space.

A product may contain multiple instances of each main or connective component. In this case, each individual element needs to be detected. Additionally, since the result of disassembly actions is nondeterministic, it is important for the system to be able to assess the success or failure of performed operations. This is referred to as *execution monitoring*. Detection of *state change*, the effect of removing a component from the product, is an essential element of execution monitoring.

Simultaneously, the vision system must be insensitive to factors which are irrelevant to disassembly, but nonetheless cause variations in the raw image data. These factors include system-related variables such as changes in ambient lighting and the positioning of the product or camera. Such problems may be resolved either at the system end (e.g. with controlled lighting) or accounted for in the software implementation. While the former may involve more detailed planning and higher setup costs, the latter is often more prone to error. In any case, slight differences in product appearance such as varied placement of stickers or the presence of dirt or rust should have minimal impact on the vision system's ability to detect and localise objects.

From a case-study of disassembly of used cars investigated by Tonko et al. [3], the authors summarise the vision-based challenges commonly encountered in the disassembly process: (a) detection of rigid objects, (b) objects located in front of a complex background, (c) partial occlusion, (d) 6-DOF estimation of objects, and (e) low contrast image due to some covering mixture e.g. oil and dirt.

Table 4.1 General requirements of a vision system for autonomous disassembly

High sensitivity	Low sensitivity
• Product (type and location) • Main components (type, location, and number) • Connective components (type, location and number) • State change	Variations in physical appearance of the same product, component or connective element, e.g. • Dust, rust, wear • Location of stickers, wires Variations of appearance originating from the disassembly environment, e.g. • Product/camera location • Sensor noise • Ambient lighting • Background

The desired qualities of a vision system for autonomous disassembly are summarised in Table 4.1.

Furthermore, since the relevant information must be gathered *on-line* during the disassembly process, execution time for detection procedures should remain minimal. Optimally, all vision processing tasks should be integrated into the process such that they are completed during the time required for physical movements, hence requiring no additional time overhead [4].

Figure 4.1 demonstrates some typical challenges that may be faced by a disassembly vision system: complicated scenes; shiny surfaces affected by background and ambient lighting; the presence of dirt and scratches; and disassembly elements such as cables and tape, which may vary in appearance between products of the same model or even during the disassembly process.

4.2 System Setup

The first step in implementing a vision system is appropriate choice of hardware. Hardware must be chosen in view of the eventual requirements of recognition and localisation. Following is a list of relevant factors influencing hardware choice.

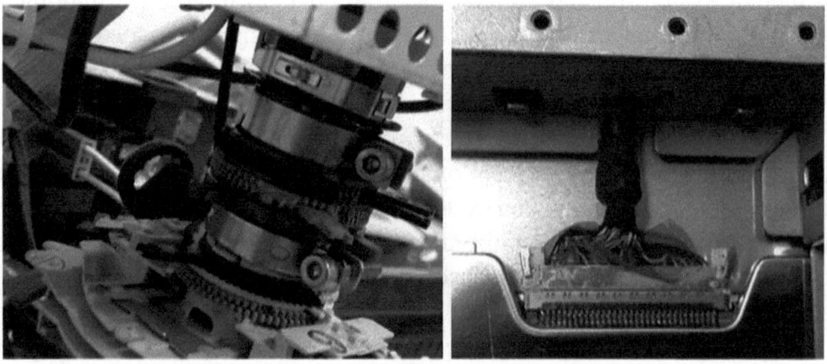

Fig. 4.1 Examples of challenges faced by a disassembly vision system

Technique and Data Collection
The fundamental technology in obtaining data controls the type and limitations of the data received. Various sensor technologies are reviewed in the following section. Furthermore, the physical setup may significantly impact the subsequent processing requirements and results gained. A camera mounted to a robot (an *eye-in-hand* camera) allows viewing of components from controlled distances and angles; however requires additional processing time for image acquisition, and has an additional error source due to robot positioning.

Resolution
The resolution describes the amount of detail in the received data. Camera resolution typically refers to the number of pixels obtained in each image. For depth cameras, the *depth resolution* (the minimum change in depth that can be sensed) may also be given, and may vary depending on distance. Since the technologies reviewed here view a scene from an approximate point, lateral resolution on objects can be increased by moving the camera closer to the object.

Sufficient resolution is required both for recognition and localisation. However, these measures of resolution do not guarantee any amount of accuracy or detail on sensed objects; noise, blur and sensor-specific weaknesses may severely reduce the useful information present in an image. Resolution comes at a trade-off with speed; image processing at excessively high resolutions only expends unnecessary time.

Range
The range of the sensor should be considered with regard to the product size and constraints in the physical setup. The field of view determines the minimum distance required between sensor and object to hold the entire object in view. The clarity of optical images also depends on focus; the focusing system inside a camera also limits the distances at which clear images can be taken. Range sensors tend to have a limited range defined in the specifications corresponding to the specified accuracy.

Frequency and Speed
Speed should be considered in terms of the entire scene acquisition process. If the disassembly system has to wait for scene acquisition, this time required directly impacts disassembly efficiency. This is primarily apparent when actuation is a part of the acquisition process, e.g. when multiple images need to be taken using one sensor at different locations to view an entire scene.

Size and Weight
The size and weight of a camera system are primarily only relevant in eye-in-hand systems, where physical size impacts the workspace and dexterity of the robot and limits its ability to undertake other tasks.

Cost
Only low-cost systems can be economically justified, due to the low gains expected from disassembly and recycling.

4.2.1 Sensor Technologies

Conventional Cameras

Similar to human's eyes, conventional cameras sense a disassembly scene by measuring the reflection of ambient lighting in the visible spectrum by the objects to be disassembled. This is by far the most common camera technology found in literature, due to their price and availability compared to other sensors, as well as the rich information that can be gained. A camera captures 2D images of a scene; multiple overlapping images taken at different locations may be used to reconstruct a 3D scene (stereo vision) at the cost of additional processing. A monochrome camera only captures light intensity; a colour camera captures the intensity at multiple wavelengths, typically red, green and blue. Sufficient ambient lighting is required to minimise required exposure time and the effect of noise. Movement during camera exposure causes blurring; hence, images should only be taken when the camera is stationary or moving slowly.

Due to the advent of the internet age, consumer webcams are now low-cost (from about $40), widely available, and offer sufficient frame rate (15–30 frames per second) and resolution for most purposes. However, the image quality is usually lower than that of high-end cameras used in specific applications, e.g. medical, surveillance camera, etc. Advantages of these specialised cameras may include higher resolution, higher frame rate, more accurate colour, less noise, less lens distortion and control of camera parameters.

Stereo vision can be implemented via the use of multiple cameras (which may be moveable or stationary) or one moveable camera. Büker et al. [5] describes a technique that reports a depth accuracy of 2 mm using two moveable cameras with controlled focus, zoom and aperture (see Fig. 4.2). Using a fixed-focus webcam in an eye-in-hand system to take multiple images along an axis, Fontes and Brandão [7] reports a depth accuracy of approximately 5 mm. In this instance the focal length was estimated in the calibration process, as such information for the webcam was unavailable. The authors attributed this inaccuracy to the focal length estimation process. Regarding recognition, difficulties result from the varied

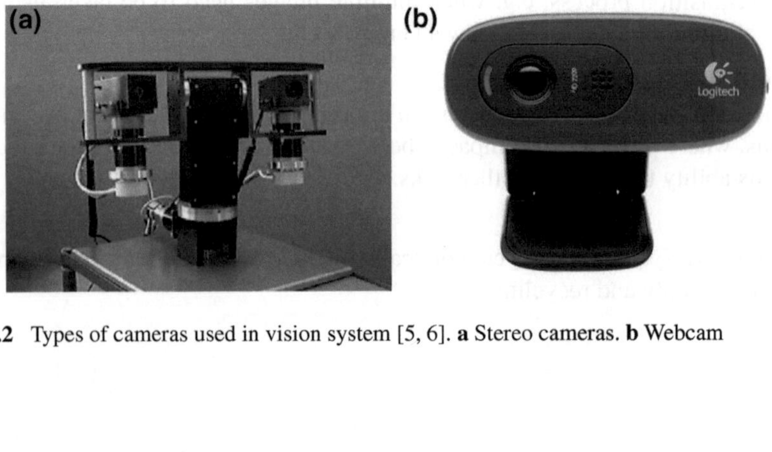

Fig. 4.2 Types of cameras used in vision system [5, 6]. **a** Stereo cameras. **b** Webcam

appearance of objects due to (a) uncontrolled ambient lighting, (b) reflection (particularly by shiny surfaces) and (c) shadows.

To minimise the effects of the ambient light, it is recommended to control lighting levels where feasible. Reflective surfaces display large alterations in appearance when the relative positions of the objects, camera and lighting are altered. Furthermore, the appearance of other objects may be reflected from a shiny surface. In our setup, we have observed that shiny surfaces reflect the orange colour of our robot. At the cost of process time, it is possible to reduce this effect by attaching a plain, neutral-coloured shield to one side of the robot and turning the robot to this side during image acquisition.

Shadows also lead to significant appearance variations in different relative positions of the objects and lighting, in particular as larger components are removed from the product. Shadowed regions have altered and potentially insufficient lighting, hindering object recognition. The effect of shadows may be reduced by using diffuse lighting sources from multiple angles, or attaching a light source as near as possible to the camera. Colour information is largely independent of lighting, however is not applicable in all contexts.

Hyper-Spectral Imaging

Information regarding the reflected spectrum of materials may aid recognition and facilitate material identification. Serranti et al. [8] describe a method for classification of recycling materials using hyper-spectral imaging in the visual and near infra-red spectrums. It was shown that it is possible to distinguish between granules of polyethylene (PE) from polypropylene (PP) in the 1,000–1,700 nm wavelength range, as well as PE + PP mix from wood, aluminium and foam in the 400–1,000 nm wavelength range. Similarly, Freitag et al. [9] report an ability to distinguish between five common types of polymers found in computer industry waste using their spectra in the 800–1,700 nm range after smoothing and differentiation. The described approaches used spectrographs with precise lighting control. A spectrograph measures the spectrum of a beam of light, precisely and at high resolution, by scattering it into its component wavelengths. However, as only one line is scanned at a time, this method requires some means of actuation and is slower than desired.

Using specialised filters atop a conventional camera sensor, one can also measure light intensity at different (albeit less precise) wavelength bands. Imec recently unveiled such a camera capable of outputting images in 32 channels in the 600–1,000 nm range at 256×256 pixel resolution [10]. This technology allows a coarse spectral analysis of the entire scene from one image, however is not commercially available at the current time. Alternatively, it is possible to measure reflectance spectra using controlled lighting, by taking multiple images while emitting different frequency light, for example from a selection of coloured and infra-red LEDs. It is uncertain whether the spectral resolution of such methods is sufficient for material classification. Additionally, such approaches may be hindered if surfaces are covered in paint, dirt or dust.

Active Depth Sensors
An active sensor takes measurements by emitting its own source of energy and measuring the environment's response. The use of an integrated active depth sensor reduces the implementation effort and computation time required for stereo vision. In comparison to stereo vision, active sensors are more resistant to changes in ambient lighting and have an advantage in localising plain surfaces lacking in features; however may be less accurate with complicated scenes where there are rapid changes in depth.

Originally developed as a gaming controller, the Microsoft Kinect [12] *uses structured light* (a projected and reflected pattern of light) to create a depth image, which is provided alongside a colour image. Due to its affordable price, integrated setup, and ease at which depth information can be obtained, this soon saw widespread application in indoor robotics. The accuracy and resolution of the Kinect is investigated in [13]. Results showed a quadratic increase in error and decrease in resolution with distance. Shiny surfaces cause areas of overexposure, leading to gaps in the depth data which are considered *blind areas*. Similar systems include the PrimeSense Carmine [11] and the ASUS Xtion PRO LIVE [14], both of which are smaller than the Kinect and require less power. In particular, the Carmine 1.09 (see Fig. 4.3) is designed for short-range sensing and may be more appropriate for automated disassembly. An external camera may also detect the projected light patterns if it is not shielded from infra-red light. This may interfere with object detection using external cameras.

A *time-of-flight* (ToF) sensor measures the time taken for a beam of light to travel the distance to and from the target. As a member of this category, 2D laser scanners have demonstrated reliable performance and are now widely accepted both in research and industry [15]. However, the need to actuate such a laser scanner to produce a 3D image again causes acquisition to be more complex and slower than desired. A lock-in ToF sensor (e.g. the MESA SR-4000) measures distance by emitting a modulated signal and measuring the phase change of the signal received at each pixel. An entire depth image is acquired in each scan, albeit at lower resolutions. A review of lock-in ToF sensors is presented in [16]. Range is limited by the modulation frequency, whereby in certain models, out-of-range points may be indistinguishable from the in-range counterpart that has a phase difference of 360°. This effect is reported to be suppressed in the MESA SR-4500 [17]. Both lock-in ToF sensors and 2D laser scanners suffer an error whereby pixels lying on edges, where emitted light is partially reflected from multiple surfaces, report false values that lie between the two depth values. Stoyanov [15] reports the results from a comparative evaluation suggested that the Kinect and SR-4000 perform (on average) equally well at "low range" (<3.6 m). The SR-4000 displayed better

Fig. 4.3 The Carmine 1.09 integrated camera and structured-light depth sensor [11]

performance when greater distances are included but was found to be susceptible to the aforementioned range ambiguity. Performance was measured by the number of (positive and negative) reference points correctly and incorrectly classified.

4.2.2 Comparative Summary

A summary comparing examples of aforementioned sensor technologies are presented in Table 4.2. The maximum values are presented for resolution and frame rate. These settings are often variable. Exposure time for cameras is required for the sensor to gather enough light; some conventional cameras or camera software may automatically lower frame rate to compensate for low lighting levels. The frame rate for ToF cameras is dependent on integration time.

4.2.3 Overview of Configuration

In this research, the vision system is designed to be low cost, while functionally meeting the requirements for automated disassembly (see detail in Sect. 4.4). The hardware was selected from the options described in previous section. As a result,

Table 4.2 Specifications and market prices of examples of aforementioned sensor technologies at the time of writing this book

Sensor	Resolution (pixels)	Range	Frame rate (fps)	Size (mm)	Apprx. cost (USD)
Webcam (logitech HD C270)	1,080 × 720	Fixed focus (manually adjustable)	30	84 × 60 × 32	$40 [6]
Webcam (logitech HD C615)	1,920 × 1,080	Autofocus	30	69 × 40 × 34	$70 [18]
High performance camera for industrial imaging (PixeLINK PL-B776)	2,048 × 1,536	Focus determined by lens (standard mount; not included)	4,000 at 8px resolution 12 at max resolution	102 × 50 × 41	$1,095 [19]
Structured light + colour camera (PrimeSense Carmine)	640 × 480	Colour camera: autofocus 1.08: 0.8–3.5 m 1.09: 0.35–1.4 m	60 30 at max resolution	180 × 25 × 35	$200 [11]
Lock-in time-of-flight sensor (MESA SR4500)	176 × 144	0.8–9 m	30	119 × 75 × 69	$4,650 [17]

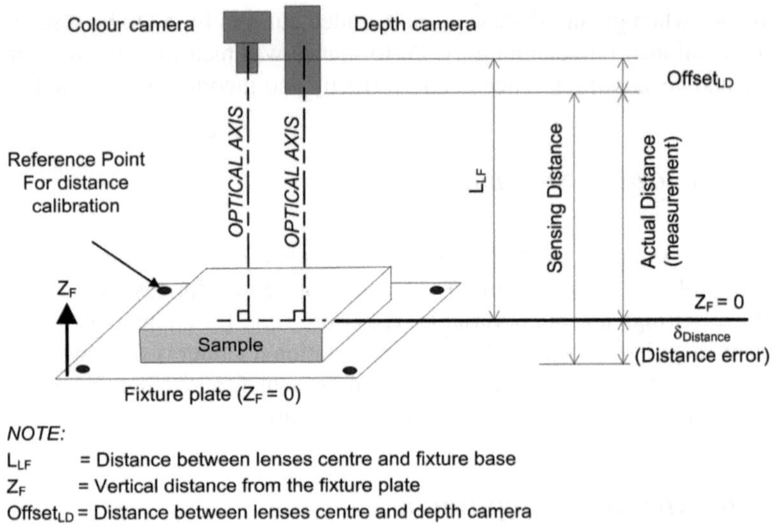

Fig. 4.4 Configuration of the cameras over the fixture plate and distance calibration

the system consists of two cameras, including a colour camera[1] and a depth camera.[2] The two cameras are mounted parallel and adjacent to each other at 1.2 m above the fixture plate which is set as $z_F = 0$ (see Fig. 4.4). The mounting location is selected to minimise the effects from lens distortion and perspective as illustrated in Fig. 4.5 (Distortion map in Fig. 4.5a is illustrated in [13]).

At this mounting distance, colour camera images can be captured with the highest precision possible (mm/pixel) while still maintaining view over the entire fixture plate (see Fig. 4.6). This position is located above the robot's working area, hence avoiding any possible crashes. Carefully adjusting both cameras to be parallel simplifies image perspective calibrations.

Controlling lighting conditions simplifies the problem of visual recognition. To reduce the effect of shading, daylight light bulbs (18 W, 6,400 K) are installed, projecting light at 45° onto the horizontal plane. Consequently, the majority of areas are lit by at least one light. High intensity light allows the colour camera to obtain better quality images with lower noise and wider depth of field. Since colour is crucial in the recognition algorithm, the setup is calibrated for colour

---

[1] *The colour camera* captures a 1,000 × 1,000 pixel single channel 11-bit image encoded with the Bayer filter, which is decoded in pre-processing. The resulting colour image is appropriate for observing small details and colour-based features. The precision in the horizontal xy-plane at $z_F = 0$ is 0.57 mm/pixel.

[2] *The depth camera* which is a Kinect sensor is used to generate 2.5D depth images. This simplifies localisation by measuring the distance to the object. The easily-obtained 3D object geometry also aids in recognition. The precision of the 640 × 480 pixel depth image is 1.74 mm in the horizontal xy-plane at $z_F = 0$ and 3.37–4.39 mm in height within the operational range.

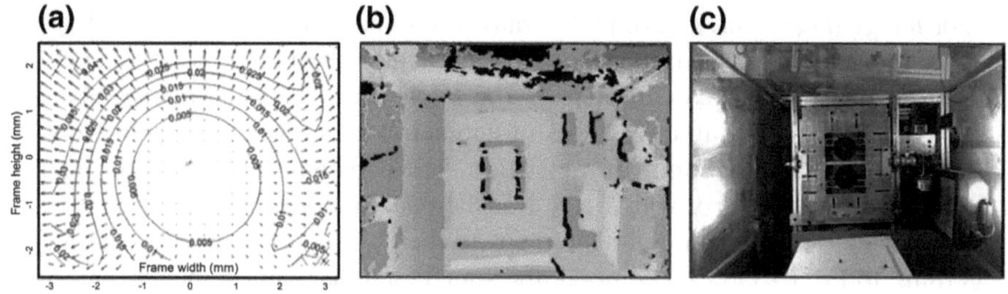

Fig. 4.5 Raw images and distortion field of the Kinect sensor. **a** Distortion of depth image. **b** Depth image. **c** Colour image

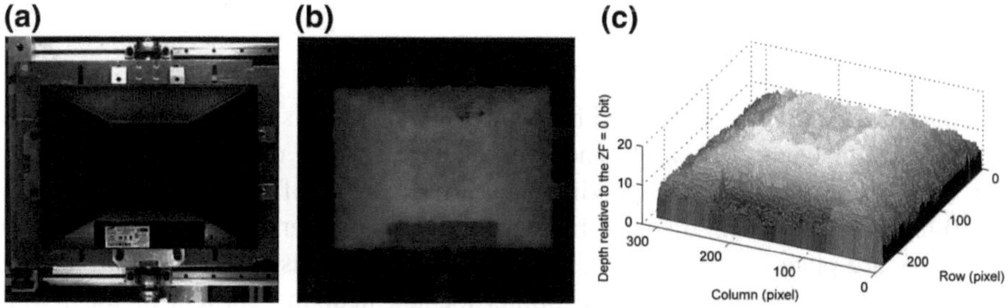

Fig. 4.6 Images from *top* view. **a** Colour image. **b** Transformed depth image. **c** 2.5D depth map

balance [18]. The hardware configuration, as well as the calibration process, are explained in detail in the following section.

4.2.4 Calibration and Localisation

Each camera senses information pertaining to an object's position with respect to its own coordinates. Information may need to be combined from multiple sources. In order to direct physical actions, the coordinates then need to be transformed into a coordinate system known by the robot. Calibration is required to discover the correct parameters for the coordinate transformations. These issues are explained with respect to our system in the following sections.

Calibration of the Depth Image

The 640 × 480 pixel depth image needs to be aligned to the 1,000 × 1,000 pixel colour image, which is the main image. The optical axis of each camera is fixed perpendicular to the fixture plate (see Fig. 4.4) so that the image planes are parallel to the fixture plate. The calibration is performed in two parts: (a) image alignment, and (b) distance calibration.

For *image alignment*, an affine 2D transformation is applied to the depth image (source image coordinates) in order to geometrically align the pixels to the colour image (destination image coordinates). The affine transformation matrix $\mathbf{M}_{Affine}$ is a 2 × 3 matrix containing three types of geometric transformation parameters: (a) 2D rotation, (b) scaling, and (c) translation. These parameters are represented as a 2 × 2 rotational matrix and 2 × 1 translational vector inside the 2 × 3 transformation matrix. The source image maps to the destination image by warping as according to Eq. (4.1) [19]. Points in the source and destination images are represented by $\mathbf{X'}_{src} = [c_{src}\ r_{src}\ 1]$ and $\mathbf{X}_{dst} = [c_{dst}\ r_{dst}]$ respectively. These locations are represented in image coordinates (c, r), where c is the column index and r the row index. The origin point $(c, r) = (0, 0)$ is located at the top-left corner of the image. The elements in $\mathbf{M}_{Affine}$ are numerically solved from pairs of corresponding points in both images.

$$\mathbf{X}_{dst} = \mathbf{M}_{Affine}\mathbf{X}'_{src} \tag{4.1}$$

Distance calibration is performed by comparing sensed distances (D_{sense}) to measured actual distances (D_{actual}). The sensed distances are calculated according to Eq. (4.2) [20] from the corresponding 11-bit pixel value of the depth image (ranging 0–2,047). The actual distance between the depth camera and the upper surface of the fixture plate is physically measured. The distance used for calibration is the average distance of four reference points located near the corners of the fixture plate (see Fig. 4.4). In this research, vertical distance is represented by z_F, the vertical distance above the fixture plate. z_F will be used for object localisation. $z_F\ (c, r)$ at a particular coordinate is computed from Eq. (4.3).

$$D_{sense}(c,r) = 123.6 \times \tan\left(\frac{PixelValue(c,r)}{2843.5} + 1.1863\right) \tag{4.2}$$

$$z_F(c,r) = L_{LF} - Offset_{LD} - D_{sense}(c,r) + D_{actual} \tag{4.3}$$

Camera Configuration and Mapping of Coordinate Frames

The relation between the image plane (spatial sampling) and operational space is determined by the coordinate mapping process. The mapping process is applied to both images after they are geometrically aligned. The frame mapping is performed based on a camera calibration matrix containing two types of parameters: (a) *intrinsic parameters,* and (b) *extrinsic parameters* [21]. The *intrinsic parameters* represent the characteristics of the lenses and the image sensor,[3] and consist of focal length (f), scale factor in both directions (α_x and α_y), and the offset of the image coordinate with respect to the optical axis (X_0 and Y_0). The *extrinsic parameters* represent the relation of position and orientation between each

[3] In this research, the approach to obtain intrinsic parameters is simplified by two assumptions: (a) the camera is equipped with low distortion lenses, and (b) the physical position and orientation of the camera are accurately adjusted.

coordinate system. Therefore, a position in the operational space can be written as a function of the image space and the aforementioned parameters as seen in Eq. (4.4), where $\{\mathbf{L}_{Offset}\}$ represents the constants involved in the translational and rotational transform between the origins of the coordinate frames.

$$position(x, y, z) = \mathrm{H}\left(c, r, z_F | \alpha_x, \alpha_y, f, X_0, Y_0, \left\{\mathbf{L}_{Offset}\right\}\right) \tag{4.4}$$

The configuration of the system is illustrated in Fig. 4.7. The system consists of four physical components resulting in four physical coordinate frames: (a) *robot base frame* {B}, (b) *fixture base frame* {F}, (c) *tooltip frame* {T}, and (d) *lenses centre frame* {L}. In addition, two virtual coordinate frames are set up to derive the geometrical relations inside the colour camera: (a) *a spatial sampling frame* {S} and (b) *an image plane frame* {I}.

In addition, the *product coordinate frame* {P} is defined in order to describe the geometry and disassembly operation parameters with respect to each product. This is the primary coordinate frame used to store product-specific information, including robot movement paths, in the disassembly process. Since the location of {P} changes according to each disassembled sample, this is not considered in the calibration phase. The conversion between {B} and {P} is done during the disassembly process.

The relation between these coordinates observed from the top-view is shown in Fig. 4.8. In summary, the coordinate frames in this system are listed in Table 4.3.

The perspective transformation in the camera is illustrated in Fig. 4.9. The spatial sampling frame defines the 2D position of a pixel on the image sensor. The

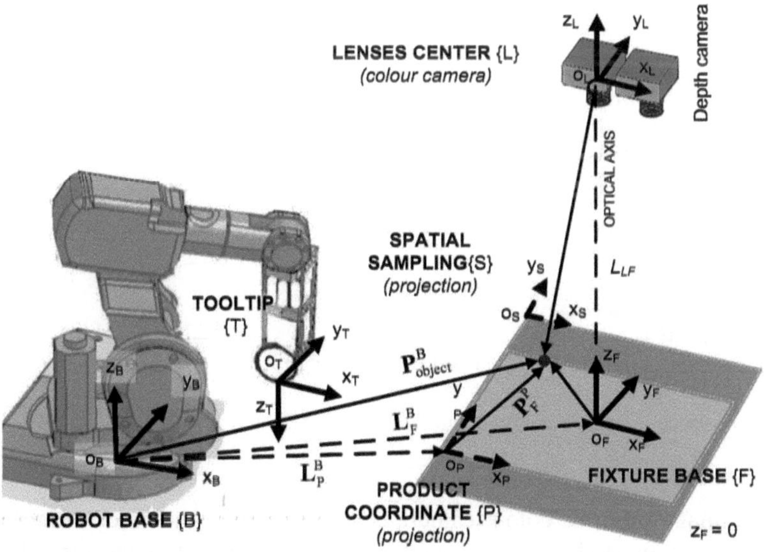

Fig. 4.7 Configuration of the disassembly cell

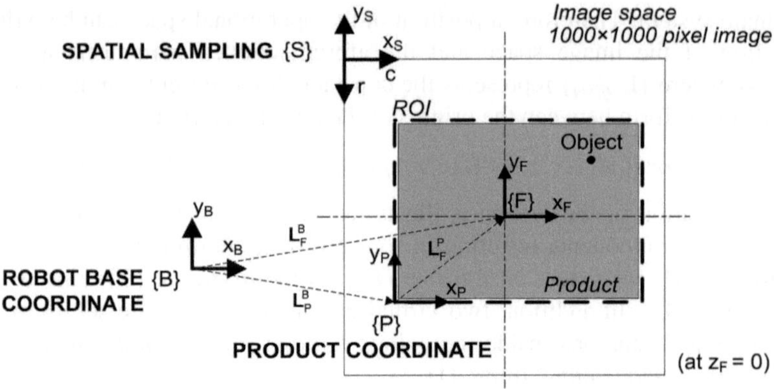

Fig. 4.8 Frames coordinate and image space observed from the top-view

Table 4.3 Summary of coordinate frames

Coordinate frame	Type	Location of the origin
{B} Robot base	Physical	Centre of robot base
{F} Fixture plate base	Physical	On fixture plate at colour camera line of sight
{T} Tooltip	Physical	End of the cutting tool
{L} Lenses centre	Physical	Centre of the colour camera's lenses
{P} Product coordinate	Physical	Bottom-left of the product sample
{S} Spatial sampling	Virtual	Top-left of the colour image
{I} Image plane	Virtual	Centre of the image sensor of colour camera

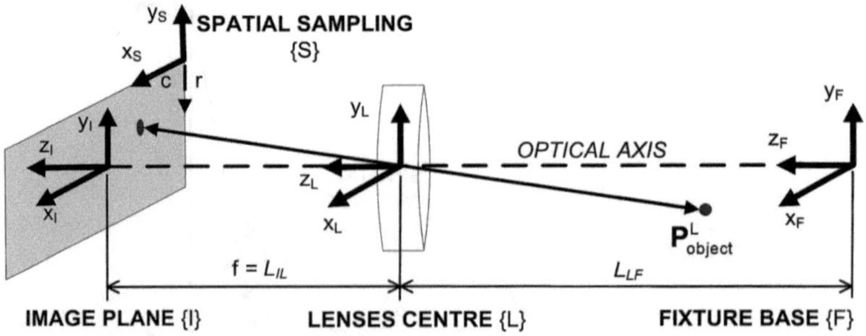

Fig. 4.9 Perspective transformation in the camera

captured image from the colour camera is a projection of the objects onto this *xy*-plane. The origin of the spatial sampling frame is located at the top left of the image. Pixel positions are represented by x_s and y_s which correspond to c and r on the image plane respectively.[4] From Fig. 4.9, the relations between the physical position of the object (P^L_{Object}) and the variables obtained from the cameras (c, r, and z_F) are presented in Eqs. (4.5–4.7).

$$c = x_S = -\frac{\alpha_x f \cdot P^L_{x,object}}{P^L_{z,object}} + X_0 \tag{4.5}$$

$$r = -y_S = \frac{\alpha_y f \cdot P^L_{y,object}}{P^L_{z,object}} + Y_0 \tag{4.6}$$

$$z_F = L_{LF} - P^L_{z,object} \tag{4.7}$$

Referring to the relation shown in Eq. (4.4), the position of the object relative to {B} can be obtained by computing the transformation matrices shown in Eq. (4.8). P^L_{Object} can be derived from Eqs. (4.5–4.7).

$$\mathbf{P}^B_{object} = \mathbf{T}^B_F \mathbf{T}^F_L \mathbf{P}^L_{object} \tag{4.8}$$

where

$$\mathbf{T}^B_F = \left[\begin{array}{ccc|c} -1 & 0 & 0 & L^B_{F,x} \\ 0 & -1 & 0 & L^B_{F,y} \\ 0 & 0 & 1 & L^B_{F,z} \\ \hline 0 & 0 & 0 & 1 \end{array}\right];\ \mathbf{T}^F_L = \left[\begin{array}{ccc|c} 1 & 0 & 0 & 0 \\ 0 & 1 & 0 & 0 \\ 0 & 0 & 1 & L_{LF} \\ \hline 0 & 0 & 0 & 1 \end{array}\right];\ \text{and}\ \mathbf{P}^L_{object} = \begin{bmatrix} P^L_{x,object} \\ P^L_{y,object} \\ P^L_{z,object} \end{bmatrix}$$

The resulting P^L_{Object} is shown in Eq. (4.9). In general, calibration can accurately determine the value of all parameters (α_x, α_y, f, X_0, Y_0, L_{LF}, $L^B_{LF,x}$, $L^B_{LF,y}$, and $L^B_{LF,z}$) using numerical methods. Alternatively, the parameters can be directly measured or obtained from the physical system, according to the aforementioned assumptions. The parameters, variables and the methods of obtaining the necessary data are summarised in Table 4.4.

$$\mathbf{P}^B_{object} = \begin{bmatrix} x_B \\ y_B \\ z_B \end{bmatrix} = \begin{bmatrix} \left[\frac{1}{\alpha_x f}(L_{LF} - z_F(c,r))\right](c - X_0) + L^B_{F,x} \\ \left[\frac{1}{\alpha_y f}(L_{LF} - z_F(c,r))\right](-r + Y_0) + L^B_{F,y} \\ z_F(c,r) + L^B_{F,z} \end{bmatrix} \tag{4.9}$$

[4] The image plane and the spatial sampling frame are equivalent concepts, with a slightly different choice of coordinate system. Image plane coordinates (r = row, c = column) are here preferred due to the definition of the image used in most image processing libraries, e.g. OpenCV.

Table 4.4 Summary of parameters and variables for calibration

Parameter/ variable	Definition	Approach for obtaining the value	Unit
L_x, L_y, L_z	Offset between {B} and {F}	Physical measurement	mm
L_{LF}	Offset between {L} and {F} along the optical axis	Physical measurement	mm
X_0, Y_0	Offset between {I} and {S}	Measurement from captured image	pixel
α_x, α_y	Scale factor	Calibrate using the measured physical size of object at $z_F = 0$ and its size on the captured image $\alpha_i = \frac{L_{LF}}{f}\left[\frac{\Delta P^B_{i/object}[\text{mm}]}{\Delta P^S_{i/object}[\text{pixel}]}\right]; i = x, y$	$\frac{\text{mm}}{\text{pixel}}$
f	Focal length of the lenses	Lenses specification	mm
c, r *(variable)*	Position on the spatial sampling coordinate frame	Output from processing the captured image	pixel
z_F *(variable)*	Vertical distance of the object from {F}	Processing from the depth image	mm

Localisation and Product Coordinate {P}

After the calibration process, the system is able to accurately determine the location of the object relative to the robot base coordinate. The locations of the object in 3D operational space are presented in two coordinate frames, {B} and {P}. The robot coordinate frame {B} describes robot movement paths within the scope of the program operated by the robot controller. The product coordinate frame {P} is used in the rest of the system. Using this frame, locations of product-specific features are described relative to the product itself. This information can therefore be used to generalise between samples of the same model or product—an appropriate frame of reference for a learning process (see Chap. 5).

The location relative to the product coordinate P^B_{Object} can be obtained by multiplying the transformation matrix $\mathbf{T}^P_B$ as seen in Eq. (4.10). According to the configuration of the system in Fig. 4.7, the final result is shown in Eq. (4.11).

$$\mathbf{P}^P_{object} = \mathbf{T}^P_B \mathbf{P}^B_{object} = \mathbf{T}^P_B \left(\mathbf{T}^B_F \mathbf{T}^F_L \mathbf{P}^L_{object} \right) \tag{4.10}$$

$$\mathbf{P}^P_{object} = \begin{bmatrix} x_P \\ y_P \\ z_P \end{bmatrix} = \begin{bmatrix} \left[\frac{1}{\alpha_{xf}}(L_{LF} - z_F(c,r))\right](c - X_0) + L^B_{F,x} - L^P_{B,x} \\ \left[\frac{1}{\alpha_{yf}}(L_{LF} - z_F(c,r))\right](-r + Y_0) + L^B_{F,y} - L^P_{B,x} \\ z_F(c,r) + L^B_{F,z} - L^P_{B,x} \end{bmatrix} \tag{4.11}$$

In conclusion, the calibration process is conducted to determine the intrinsic and extrinsic parameters of the system. A careful consideration of physical system design can reduce the number of calibration parameters and therefore simplify the calibration process. The disassembly process operates according to positions sensed in 3D space. These positions can be calculated using the calibration parameters and variables c, r, and z_F which are directly obtained from the images. The product coordinate {P} is used in majority of the process since it represents the geometry of the product itself. However, the robot coordinate frame {B} is also necessary in order to control the movement of the robot.

4.3 Recognition Techniques

The human vision system is excellent at recognising arbitrary objects. A myriad of work has been done in object recognition, yet the current best solutions in computer vision do not compare with the speed and accuracy of our own vision. The problem remains an active research area; this section only seeks to provide a very brief overview of the topic with specific emphasis on prior work in disassembly vision. The problem of recognition roughly consists of the steps of (a) methods of extracting and describing features, and (b) classification.

A *feature* is any measure that can be obtained from the raw data, used to identify objects of interest, e.g. a pixel (colour/lightness intensity), a location, an edge, a corner or any combination of such measures. Some method of *classification* then decides whether or not a region in the image constitutes an instance of the object of interest. At some point, regions of the image may also be *segmented* into different regions to be separately analysed. At least one known instance of the object of interest is required to allow comparison. A larger sample size is typically used to account for natural variations in appearance.

The appropriate choice of techniques largely depends upon application context. Common techniques, particularly those previously used in disassembly robotics, are described in the following sections.

4.3.1 Thresholding

An image obtained from a camera is represented as a series of numerical values signifying the intensity of light received at each pixel, i.e. from a particular direction. *Thresholding* is the conversion of such an image into one where each pixel only has a binary value by using some kind of limit or threshold.

A colour camera measures the intensity of light in multiple wavelengths. This information can be described in various *colour spaces* depending on the required functionality. The RGB colour space describes the representation in terms of light intensity in the red, green and blue channels respectively. The Hue-Saturation-Value (HSV) colour space is commonly used to isolate the chromatic properties from brightness. Colour can thus be considered independent of ambient illumination [22]. Converting between colour spaces is simply a mathematical transformation of the channel values for each pixel.

Where the object or background is of a known brightness or colour, the pixel values themselves may be used as the primary method of classification. A threshold colour may be set: pixels with values lying on one side of the threshold are considered to belong to the object of interest; those on the other, as the background. The object locations are then extracted from the resulting binary image. Adaptive thresholding is where the threshold is not constant, but rather derived from the properties of the image or regions within the image.

Thresholding is commonly-used in human skin tone detection [24]. Multiple instances of objects detected through thresholding can further be grouped into regions of conjoined or nearby pixels (see Fig. 4.10). This falls under the area of *clustering* or *blob detection*. Thresholding is also common as a pre-processing technique (e.g. [25]) to simplify the image for further processing.

Thresholding is also useful in the interpretation of depth images. Background clutter can be automatically rejected if it is known that the object lies within a specific height or distance. Similar to colour blob detection, objects can also be segmented using the criterion of minimum depth change or distance from other objects.

4.3.2 Edge Detection and Contour Geometry

Contours or ***edges*** are lines separating regions of contrasting colour or intensity. On a camera image, edge detection is commonly performed by approximating the derivative of the image (*image gradient*), which has greatest magnitude in regions of intensity change. The direction of the image gradient can be used to determine the direction of the edge. The *Canny edge detector* [26] (see Fig. 4.11) builds on these principles in a technique that outputs a binary image where edges appear as single-pixel-wide lines.

The shape of an edge is naturally descriptive of the object and its geometry, however this data must be further processed into a form that can be compared or analysed. *Hough transforms* locate known and geometrically-defined objects within an image by considering each point as a vote towards all valid transformations of the object in which the point can participate. Votes are tallied in discrete

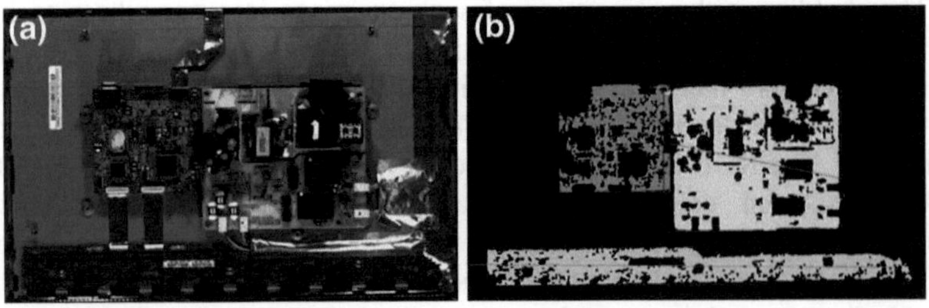

Fig. 4.10 Thresholding and blob detection for the detection of PCBs [23]. **a** Captured image. **b** Blob detection on morphed positive area

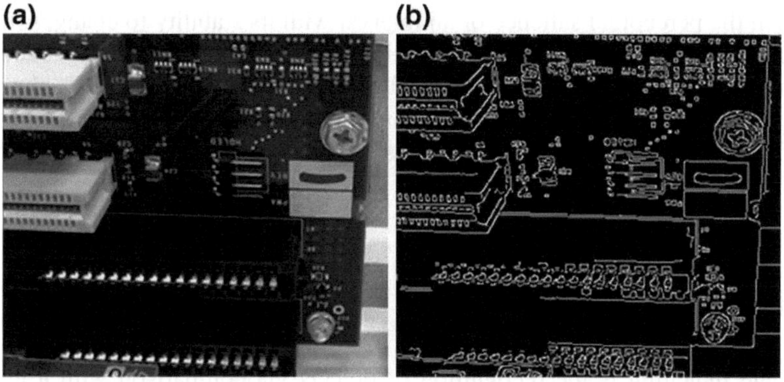

Fig. 4.11 OpenCV Canny edge detector applied to an image of a PCB. **a** Captured image. **b** Canny edge detection

bins. Actual occurrences of the objects within the image cause local maxima in the tallies, from which the parameters of the representation (e.g. distance and angle for a straight line) can be directly obtained.

The *Ramer-Douglas-Peucker* or *split-and-merge* algorithm approximates the representation of a line of unknown shape to a series of segments. If the distance between a point and the proposed segment is greater than a threshold, the line is split and a shorter segment extending only to the problematic point is proposed, continuing until all points in the line lie within the given tolerance to the fitted segments.

Using a depth image, [27] uses edge detection to recognise grasping surfaces for a parallel gripper approaching the objective module in a camcorder from above. Parallel, straight edge sections are found after applying the split-and-merge algorithm using circular segments. Candidates are subsequently analysed to ensure the edges belong to the same object and provide enough space for the approach of the gripper.

Inspired by the human vision system, Büker et al. [5] uses Gabor filters to recognise contours at various resolutions and orientations. A General Hough Transform is then used to determine the location of nuts. This method is robust to background clutter, however the presence of dirt and rust meant that nuts "did not exhibit enough contours to determine their position accurately".

After contour image extraction, Gil et al. [25] uses a Progressive Probabilistic Hough Transform to estimate straight-line sections of wires, from which cutting positions can be estimated. The split-and-merge algorithm using straight-line segments was proposed for the detection of screws. In the displayed results an edge detector appears to have been applied twice. Geometric features of edges, namely lengths, orientations and their distributions, as well as template matching, are used in component recognition.

Scale space filtering analyses edges and intervals (the areas separated by edges) in different *scales*, i.e. after different amounts of smoothing. Considering edges as the zeros of the second derivative, Witkin [28] observes a correspondence between the perceptual salience of an interval with its stability to changes in scale. A "top-level" description or segmentation of the intervals can be obtained by finding the first local maxima in stability with decreasing scale (as the image becomes increasingly detailed). An edge found at a coarse scale can be localised at a finer scale. Hohm et al. [29] suggests scale space filtering as an appropriate means for the detection of cables.

4.3.3 Template Matching

Template matching is the recognition of objects via comparison with a template (i.e. example or model image). The simplest form is pixel-by-pixel comparison of regions within a scene to the template. This may be performed on the raw image or after pre-processing (e.g. edge detection). Template matching is sufficient provided that the objects of disassembly are known and the distance and orientation of the camera are either maintained or the variations compensated for. To reduce the number of comparisons required, it is common to *subsample* the image, i.e. perform the comparison on images of reduced resolution. *Image pyramids* are typically used as an efficient method for subsampling.

This is the method described in Elayed et al. [4] to recognise components within a personal computer after "appropriate scaling, rotation and histogram-equalization" following gross scene segmentation using a depth image. Following the localisation of ceiling beams using a Hough Transform, Rolando Cruz-Ramírez et al. [30] generates a new trajectory for the acquisition of images, maintaining a fixed distance between camera and beam. Template matching is then used to generate hypotheses for the locations of screws. In both aforementioned examples, knowing the location of the containing structure limited the required search space. For component recognition, Gil et al. [25] uses template matching aided by image pyramids, whereby candidate locations are initially found on a low-resolution image. These potential matches are then verified and refined at higher resolutions.

4.3.4 Keypoint Feature Matching

Keypoints are distinctive locations within an image which can be computed and used to recognise objects in a scene (see Fig. 4.12). A keypoint detector extracts locations of interest that optimally occur at repeatable positions in images from different viewpoints, and the descriptor aims to provide a concise and easily-comparable description of the area around each location. The keypoints, instead of every individual pixel, are then compared, often reducing the number of total comparisons that need to be made. Provided that enough keypoints exist on the object, such methods are more efficient than direct template matching when the scale and/

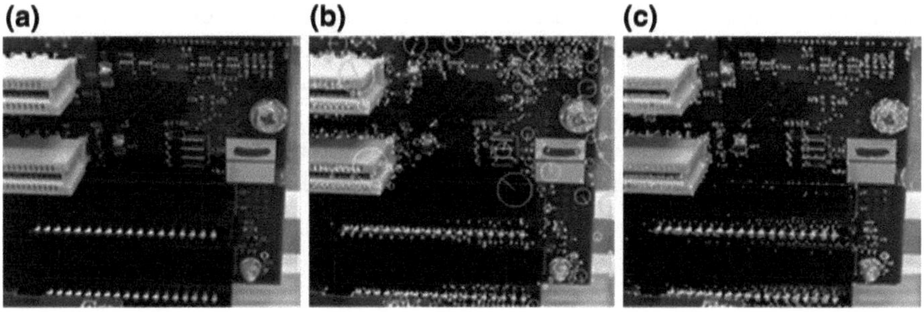

Fig. 4.12 Keypoint detection using SIFT and FAST algorithms. **a** Captured image. **b** SIFT. **c** FAST

Table 4.5 Table of common keypoint detectors and descriptors [31–39]

Blob detectors	Corner detectors	Other descriptors
• SIFT* [34]	• Harris* [38]	• BRIEF [40]
• SURF* [35]	• FAST [39]	• FREAK [41]
• CenSurE/STAR [36]		
• BRISK* [37]		

or orientation of an object are uncertain. Examples of popular keypoint detectors and descriptors are given in Table 4.5. Blob detectors are typically more stable (robust to changes in viewpoint) than corner detectors; however, corner detectors have the advantage of speed and greater accuracy in localisation [31].

4.3.5 Semantic/Relational Features

Humans recognise objects, not only from the appearance of each object alone, but rather, in context of the scene and prior knowledge. External information, such as size and location with respect to other components, may produce constraints for where objects can be found, and aid in disambiguating objects of similar appearance. A survey of general approaches to obtain and exploit context-based visual information in computer vision is found in Galleguillos and Belongie [40].

As opposed to general image classification, proposed disassembly systems operate in constrained contexts. In such contexts, it is common to first localise the product to be disassembled, then restrict the search space for smaller components to the area encompassing this product [4, 30]. Jørgensen et al. [41] uses a hierarchical tree model of products, whereby each node represents a component and the root nodes represent product classification. Nodes contain information both regarding normalised expected product position and a weighting factor based on the probability of occurrence within the product, which are used in determining the best identification for detected components. Karlsson and Järrhed [42] describes a vision system for the disassembly of electric motors, whereby the regions of interest (ROI) for screws are taught manually by a human operator for every model. In future appearances of these models, only the ROIs are searched to verify the presence of screws. The existence of screws in the pre-learned locations is then used to confirm the identification of the motor. For bolt detection on car wheels, Büker et al. [5] further specifies that the bolts lie in a symmetrical arrangement; this is integrated into their Generalised Hough Transform approach.

4.3.6 Further Classification Methods

It is difficult to determine the exact qualities or feature values defining the identification of an object. Statistical and machine learning techniques are commonly used for the classification of a new instance with respect to a known data set. Following is a brief overview of some such techniques.

Fuzzy measures allow measurement of the correspondence of a new instance to a template by first calculating the *membership* or agreement of each measured feature to those of the template. In order to identify incoming products (electric motors), Karlsson and Järrhed [42] manually selected a set of 8 features calculated from a binary image containing the silhouette of the product: area, perimeter, moments of inertia I_{xx}, I_{yy} and I_{xy} and the measurements of the bounding rectangle (long side length, short side length and side ratio). Initial testing showed Gaussian-shaped fuzzy measures to be most appropriate. Each feature in each template (i.e. from one motor type) is assigned a Gaussian membership function, characterised by its mean and standard deviation. This measures how well a new instance correlates with the template according to this individual feature. To produce a fuzzy measure for the template, feature membership values are then fused using Eqs. (4.12) and (4.13).

$$f(\mu_1, \ldots, \mu_n) = \left(\frac{1}{K^N - 1}\right)\left(\frac{-1 + K^N G}{1 + G}\right) \tag{4.12}$$

$$G = \prod_{i=1}^{N} \frac{1 + (K - 1)\mu_i}{K - (K - 1)\mu_i}; \quad 1 < K < \infty \tag{4.13}$$

μ_i = membership to feature; N = number of features; K a constant parameter [42].

Artificial *neural networks* are methods for function learning and approximation, inspired by biological systems of neurons, often used for classification using multiple inputs. Jørgensen et al. [41] proposes a component recognition system consisting of multiple pre-processing filters (including adaptive thresholding, histograms and edge and corner detectors), feature selection using a "principle component like algorithm", and input of selected features into a RAM-based neural network used for component recognition. Each neuron in a neural network is simple function with multiple inputs. These outputs can further be used in successive levels of neurons, allowing approximation of more complex functions. In order to obtain the parameters of the neuron functions, the neural network must first be trained with examples.

The *support vector machine* (SVM) is a similar technique whereby each instance for classification is considered as a point in multidimensional space defined by the value of each feature. Classification occurs according to a function (of defined type) that separates positive and negative training examples by the largest possible margin. After finding template matching alone to be insufficient for the detection of low-resolution screws, Rolando Cruz-Ramírez et al. [30] obtains greater accuracy by further assessing candidates using a SVM classifier.

Decision trees are a machine learning technique whereby the value of individual features are chosen as branching criteria which are successively applied to classify the object of interest. *Boosting* allows more accurate classifiers to be trained by using the votes of multiple weaker (less accurate) classifiers. For object detection, Viola and Jones [43] uses the boosting algorithm AdaBoost on Haar-like features, coarse features which are efficiently calculated and describe the relative brightness of adjacent regions within an image. These features are easily scaled for comparison with a scene image, however are too numerous for comparison of all features. Small classifiers are trained to provide high detection rates, which, during classification, are applied in consecutive stages (i.e. as a *cascade*) to quickly reject windows that are not of interest. For facial recognition, this method was shown to be comparable to existing literature in accuracy, with a significant increase in speed such that it is suitable for real-time applications. However, a large training set (perhaps a few thousand positive and negative labelled images) is required to train such a machine learner, making this method only suitable for standardised and commonly-appearing components.

4.3.7 Conclusions

There currently exists a wide selection of tools and techniques for the purpose of object recognition in computer vision. It is difficult to compare the performance of the various methods in the disassembly vision system due to the variety of tasks, including the image quality and difficulty of the tasks, and system setups described. Additionally, many publications lack information important for the well-informed implementation of an image processing system. Such information includes (where relevant) the algorithm used and the relevant parameters, image resolution, true and false positive rates, accuracy and execution time (with respect to the computer system).

There is so far no general solution for the recognition of disassembly objects, due to the broad scope of the required tasks. Solutions reporting positive results typically employ a range of techniques in succession, and exploit information regarding the relationship between objects. The choice of features needs to be suited for the task, such that (a) the features are capable of making the necessary distinctions and (b) the number of necessary comparisons does not lead to prohibitive execution time. In order to achieve this with commonly-available technology, and as is the pattern in literature, we suggest an approach whereby the disassembly vision system is considered as an amalgamation of separate tasks, according to different levels within the product. In this way, the requirements of each task can be specifically addressed, and specific characteristics exploited.

4.4 Requirements and Functionality

A *flexible* disassembly vision system requires its functions to be applicable to a large number of product models, not only those that are already known, but with the potential to easily integrate future models. However, constraining problems to specific applications is beneficial for reducing complexity and improving the robustness of a vision system. It is recommended to separate the problem into tasks which are as general as possible with regards to disassembly object, yet limit the scope and requirements of each individual task. This book hence considers the four following categories: (a) *product detection*, (b) *main component detection*, (c) *connective component detection*, and (d) *state change detection*. Their respective requirements and characteristics are detailed in the following sections.

4.4.1 Product Detection

The product is the largest and highest-level object that needs to be detected in a disassembly vision system. Segmenting the product from the background can be simplified by controlling the background colour (or lighting the background to take a silhouette image), thresholding by depth or location, or determining the difference between the occupied and empty rig. High resolution is generally available: general dimensions and shape can generally be obtained in high detail; however the challenge is to generate a sufficiently small set of features for quick comparison with a potentially-large database of previously-seen products.

A given product has a standard set of components, however may be altered due to repairs and maintenance. When the components are fairly standard, these, or the location of fastening elements may be used to confirm or reject hypotheses for product identification. When multiple products have a similar outer appearance, or multiple possible internal configurations exist for the one product, it may be possible to begin disassembly with multiple possibilities on hand. The ambiguity can be automatically cleared once more of the internal structure is revealed. Further product-specific features that may aid in recognition are identifiers such as logos, text and barcodes. However, Kopacek and Kopacek [44] reported that the barcode itself is insufficient for the full identification of mobile phones.

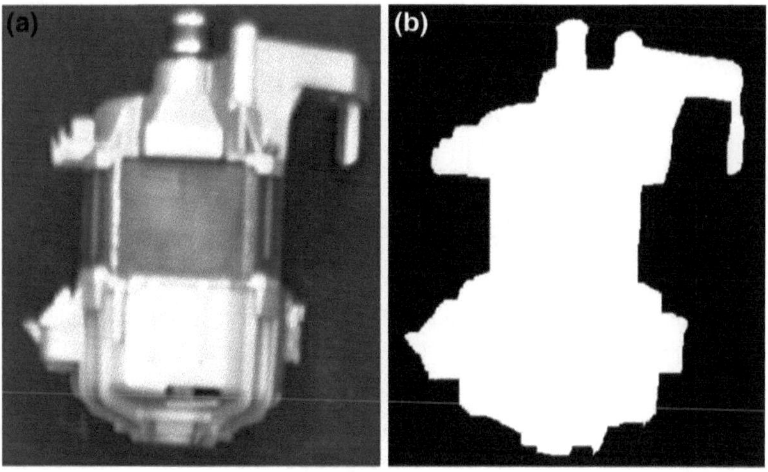

Fig. 4.13 Image processing for motor model identification [42]. **a** Captured image. **b** Processed image

Using controlled background and lighting, thresholding and morphological operations, Karlsson and Järrhed [42] obtains the binary silhouette of electric motors for disassembly (see Fig. 4.13). From these shape profiles, the values of 8 hand-picked geometric features are calculated and compared to the database using fuzzy measures (see Sect. 4.3.6). The screw locations are further used to confirm the classification. This method reported 95 % accuracy from two perpendicular viewpoints, leading to a total accuracy of approximately 98 %.

In this research, SURF is used to recognise the model of LCD screen samples. SURF keypoint features are extracted from the image of the back cover and compared with that of candidate models in the database. The model is recognised if a sufficient number of keypoints is matched. From a preliminary test on 37 different models of LCD screens, this detector achieves 95 % accuracy in which the screens were classified by the threshold of 15 % of keypoints matching. For the same models of screen, keypoints matching varied between 17–100 % according to the noise and little different in lighting condition. An example is shown in Fig. 4.14.

4.4.2 Main Component Detection

The problem of component detection can be approached from two perspectives: (a) identifying the component by *model* or (b) identifying the component by *type*. The former is more appropriate when the exact component models are known, for example when the goal of disassembly is to obtain specific pieces for reuse. In this case, a strategic use of template matching is sufficient [4, 25]. Alternatively, the goal of component detection may be to recognise components by *type*. Products within the same product family tend to contain the same types of components,

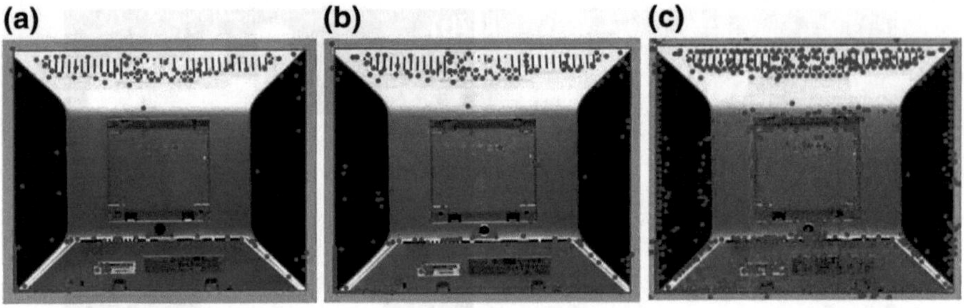

Fig. 4.14 SURF keypoints detected in a sample and a candidate model. **a** Sample. **b** Matched keypoints. **c** Detected keypoints

however with varying physical appearances, due to varying specifications and designs. If *types* of components can be recognised, the system is theoretically able to disassemble even unknown models, with the specific goal of treating each component appropriately. This is more appropriate for recycling and waste treatment. However, a more flexible classification technique is required. In this book, the concept of *common features* for detecting main component type is proposed.

The *common features* of a component type are physical characteristics which are shared between most or all components of this particular classification. Despite design differences, models within a component type often share consistent similarities, as particular features are directly related to the functionality of the component. These common features are used to formulate the rules for component recognition, which in this book will be expressed using logical notation. It is assumed that a particular component is detected at the current state if the corresponding set of component-specific rules is satisfied. The format of the general rules is shown in Eqs. (4.14) and (4.15), which state that an object x is a component type y if the object x satisfies all rules corresponding to the component type y.

$$\left[rule_1(x,y) \wedge rule_2(x,y) \wedge \cdots \wedge rule_n(x,y)\right] \supset component(x,y) \tag{4.14}$$

$$component(x,y) \equiv \left[object(x) \wedge componentType(y) \wedge (x \in y)\right] \tag{4.15}$$

The accuracy of detection depends directly on the number of significant common features and how well they distinguish between the various components. A significant number of samples need to be observed in order to define the component types, select the common features and define the appropriate parameters for classification in view of the capability of the vision system. The algorithm to calculate the truth value of each abstract rule also needs to be developed.

The most appropriate set of common features differs according to product type. The following methodology demonstrates the application of this concept to the detection of components in LCD screens, the case study presented in Chap. 6. 37 different models of LCD screens have been observed based on the proposed camera configuration. The common features can be categorised into three main groups: (a) *geometry*, (b) *colour range*, and (c) *texture and connected regions*.

Geometry
The geometry of components can be observed directly from the colour and the depth images. The size, location and shape of an observed component can be roughly described by its *minimum bounding box* (MBB), the smallest box which contains the entire component. MBB size, aspect ratio and height (distance from the front of the LCD screen) are observed to fall within consistent ranges for each type of component. The corresponding rule is then formulated by giving the possible range according to these properties. This rule is shown in Eq. (4.16) where *prop* can be substituted for size, aspect ratio, and height.

$$rule_{prop}(x,y) \subset [component(x,y) \wedge \{prop_{\min}(y) \leq prop(mbb(x)) \leq prop_{\max}(y)\}] \tag{4.16}$$

A good example for using geometry as a distinguishing common feature is the LCD module, which largely determines the size of the product as a whole. Parameters for this rule can be obtained from standards [45]. The diagonal size of the LCD modules to be disassembled in the disassembly rig ranges from 15" to 19". The height of LCD modules ranges between 10 and 20 mm and their aspect ratio between 4:3 and 16:10.

Colour Range
Due to functional material specifications and production techniques, only minor colour variations are found in most component types. It is assumed that the component is detected if a sufficiently large region of connected pixels within the corresponding colour range is detected. Pixel colour is expressed using the HSV colour space [46]. Blob detection [47] is used to locate regions of interest, that both satisfy the colour criteria and have area greater than Φ_{Blob} [see Eqs. (4.17–4.18)]. If x is a component of type y, the colour criteria (associated with h, s, v) is met for pixel I belonging to the area of blob detected in x if it has both hue and saturation values within the range defined for component y.

$$\begin{aligned} rule_{colour}(x,y) \subset \big[& component(x,y) \wedge \{area(blob(x)) \geq \Phi_{Blob}(y)\} \\ & \wedge satColourPixel(I,h,s,v,x,y)\big] \end{aligned} \tag{4.17}$$

$$\begin{aligned} & satColourPixel(I,h,s,v,x,y,) \\ & \equiv \big[component(x,y) \wedge pixel(I, colour(h,s,v)) \\ & \quad \wedge \{H_{\min}(y) \leq h(x,I) \leq H_{\max}(y)\} \wedge \{S_{\min}(y) \leq s(x,I) \leq S_{\max}(y)\}\big] \end{aligned} \tag{4.18}$$

Colour range can effectively distinguish printed circuit boards (PCBs) from other main components in LCD screens. PCBs are generally green or yellow, while other components are gray. The gray colour is due to non-colour-coated metal. Parts made of hi-strength and thick plate steel, e.g. carriers, are matte gray. Parts used purely for covering purposes, e.g. the back of the LCD module and PCB cover, are made from light plate steel which is light gray. Significant difference in H and S channels can be clearly shown as histograms in Figs. 4.15 and

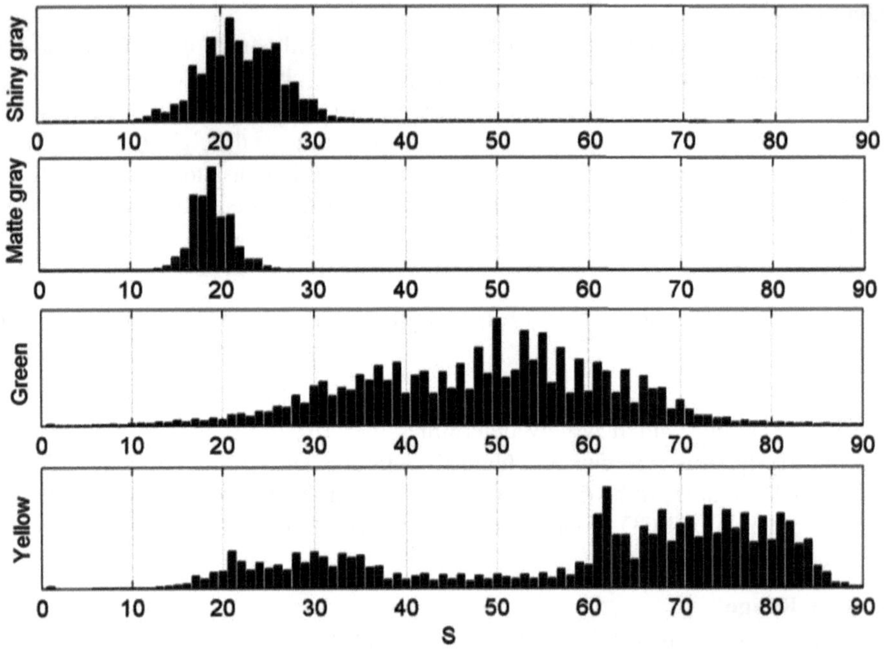

Fig. 4.15 Histogram of the base colour in the S-channel as collected from the samples

4.16 where the colour pixels are collected from the observed samples. The colour range of each component is summarised in Table 4.6. Classification between the components is done using fixed-level thresholding.

Texture and the Connected Region

The observed surface texture is also directly related to the function of the component. This can be classified into two categories: (a) *homogeneous* and (b) *non-homogeneous*. *Homogeneous* surfaces are large connected regions with minor variations in colour and depth due to small features such as ventilation holes. These are typically observed on the metallic main components, e.g. PCB covers, carriers, and LCD modules. Components having *non-homogeneous* texture, e.g. PCBs, usually contain sub-components which are noticeable due to their distinctive colours and depth variations from the base.

A connected region is recognised as a component using the rule in Eq. (4.19) associated with two key indicators: (a) *homogeneity* and (b) *significant size*. The *homogeneity* of the detected area is determined by the ratio between the blob cluster and the size of its minimal bounding rectangle (MBR) [see Eq. (4.20)]. The MBR is used as the region is considered in 2D. The MBR must have significant size in comparison to the area of the entire product. This can be determined by Eq. (4.21). The threshold Φ is used to justify both rules.

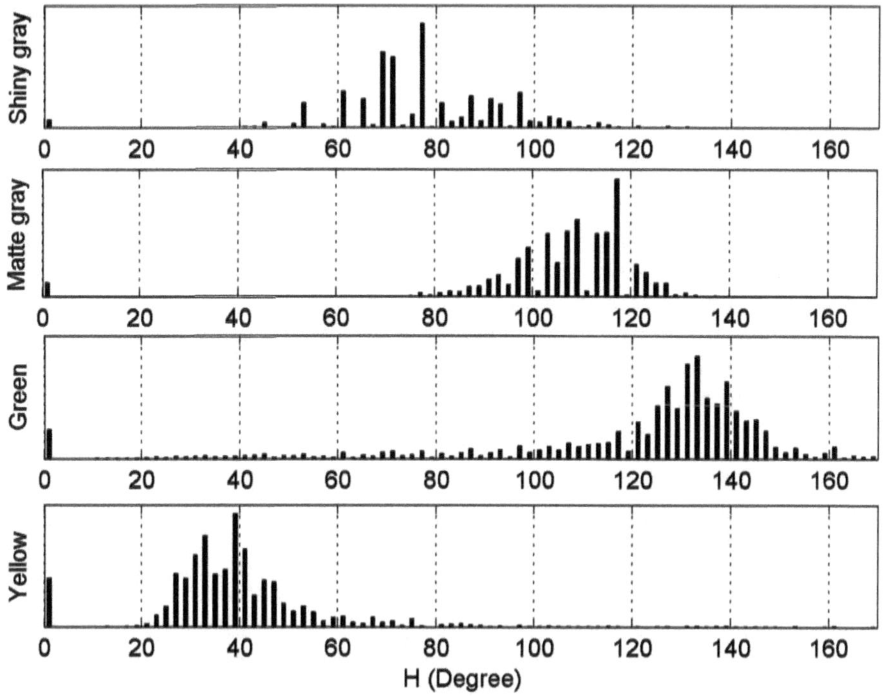

Fig. 4.16 Histogram of the base colour in the H-channel as collected from the samples

Table 4.6 Colour ranges of the components in LCD screens

Component	Colour name	HSV colour range			
		H (0,360°)		S (0,100)	
		Min	Max	Min	Max
Back cover	n/a	–	–	–	–
PCB cover	Matte gray	73°	135°	10	27
	Light gray	40°	128°	9	35
Carrier	Matte gray	73°	135°	10	27
PCBs	Green	70°	200°	35	80
	Yellow	20°	70°	35	90
LCD module	Light gray	40°	128°	9	35

$$rule_{connectedArea}(x, y) \subset component(x, y) \wedge homogeneity(x, y) \wedge significantArea(x, y) \tag{4.19}$$

$$homogeneity(x, y) \equiv \frac{area(blob(x))}{area(mbr(x))} \geq \Phi_{Blob/mbr}(y) \tag{4.20}$$

$$significantArea(x) \equiv \frac{area(mbr(x))}{area(Product)} \geq \Phi_{Mbr/Product}(y) \tag{4.21}$$

In addition, the surface roughness rule in Eq. (4.22) holds if the surface roughness value (R_a) as measured from the depth image lies within the maximum acceptable roughness for the corresponding component ($R_{a,Max}$). The capability of the depth camera must be taken into account in determining this threshold.

$$rule_{roughness}(x, y) \subset \left[component(x, y) \wedge \left\{R_a(x) \leq R_{a,Max}(y)\right\}\right] \quad (4.22)$$

In conclusion, this section explained the principle of using common features for detecting main components in disassembly. The detection rules can be formulated from the common features and parameters obtained by observing a number of product samples. An example of a set of rules and common features used to classify between types of main components was shown using the case-study of LCD screens. Further details regarding the detection of main components in LCD screens are given in Sect. 6.6.1.

4.4.3 Connective Component Detection

Automatic detection of connective components, or *fasteners*, is highly desirable in a disassembly vision system, both to verify the existence, type and location of the fasteners before executing an operation, and for the potential of executing these fundamental operations on unknown product variants. However, this is at best a challenging task due to their small size and the varying appearance of connections. Table 4.7 qualitatively lists common fasteners by their ease of detection and disassembly [48]. Due to the significantly-varying nature of the different fastener types, detection of different categories should be considered as separate tasks within the vision system.

Ease of detection was evaluated on the basis of the standardisation of the components and their appearance, as well as their general visibility. Snap fasteners, welds and glued joints are non-standard and commonly hidden from view. Automatic detection of these appears infeasible for current disassembly automation. In this case, one can only resort to using databases of known products, case-by-case teaching or bulk destructive disassembly. On the other hands, screws, bolts and rivets are of standard geometry, and are easy for a human to recognise. Screws and bolts may be lost or exchanged during product maintenance.

Table 4.7 List of common fasteners by ease of detection and disassembly with their best practice

	Disassembly method		
Detectability	Non-destructive	Semi-destructive	Destructive
High	Screws, bolts, etc.	Rivet	–
Medium	Spring, staple, cotter pin	Solder, cable, bundler, tape	Seam
Low	–	–	Snap fit, weld, glue/seal

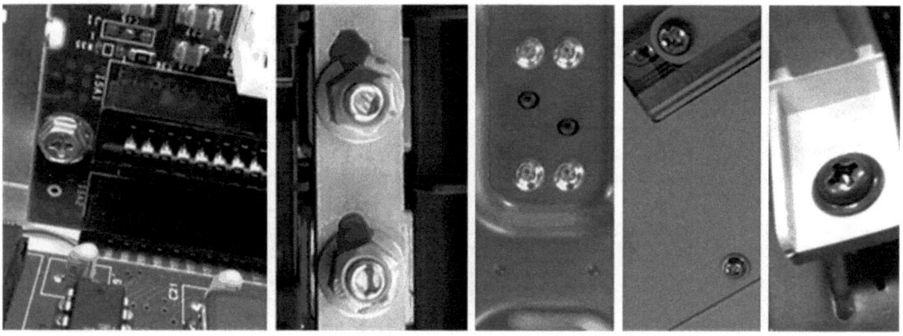

Fig. 4.17 Various screw-type fasteners found in end-of-life products

Cables, bundlers and tape are generally visually salient; however pose a challenge due to their lack of fixed shape. Springs, pins and staples have a more defined shape; however tend to be less distinguishable from other components. Due to their common occurrence within electronic products, the detection of screws and cables will be examined in more detail.

Screws, Bolts, and Nuts

Screws, bolts and nuts are defined by their shape, being typically circular with a particular standard pattern that forms surface contact with a tool (see Fig. 4.17). The main challenges posed by the detection of screws and bolts are their small size and often shiny surface. Due to their small size, either a high resolution camera or a movable camera (e.g. mounted on a robot) is required to obtain the raw images. The detail of the screw head is generally undetectable using currently-feasible active depth sensors. The shiny surface means that greater variation in appearance is expected from the reflection of the lighting, background and other components, as well as surface variations. As such, a method is required that is either edge-based and robust to extraneous salient edges (caused by varying light reflections caused by altered location/orientation or screw head shape); or very fast, enabling comparison to a large database of known examples. Unlike product and component detection, screw detection must be insensitive to changes in colour; such fasteners are made from a range of materials and colours, and may change colour due to rust.

Encouraging results for nut and screw detection were reported in [5], who reported detection a rate of 98 % for screw detection. Rolando Cruz-Ramírez et al. [30] also suggests that high detection rates may be attainable after the integration of multiple images taken from different perspectives. Both approaches demonstrated a strategic use of contextual information, first finding the containing element (the car wheel [5] or the ceiling beam [30]), then using this information to selectively take higher resolution images for the detection of the small fasteners. Figure 4.18 shows the detection of bolts in [5], whereby both the shape of the bolt and the symmetrical geometric arrangement is used, hence ruling out the multitude of other candidate locations that fit the shape criteria alone. The accuracy in [30]

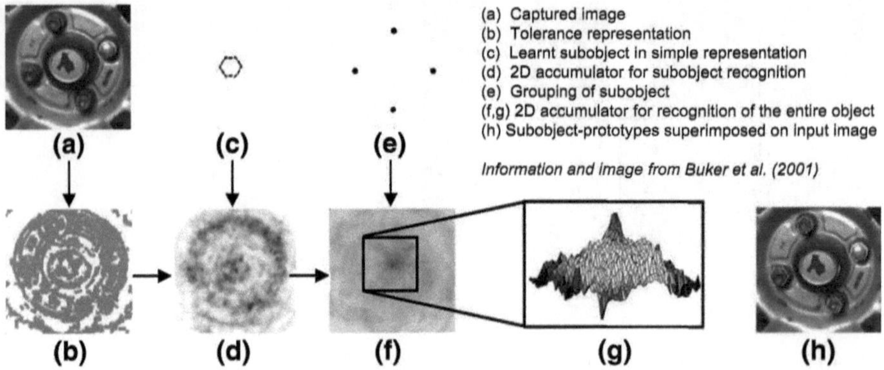

Fig. 4.18 Bolt detection, presented by [5]. **a** Captured image. **b** Tolerance representation. **c** Learnt subobject in simple representation. **d** 2D accumulator for subobject recognition. **e** Grouping of subobject. **f** 2D accumulator for recognition of the entire object. **g** 2D accumulator for recognition of the entire object. **h** Subobject-prototypes superimposed on input image

is reported for "temporal multi-image integration": 30 images are taken per metre along the ceiling beam, and the data from multiple images fused to produce the final result.

Cables

Cables are generally salient within an image due to their contrasting colour, however are flexible and only fixed in specific locations, being otherwise free to move. Cables are therefore expected to change shape during use and disassembly; such changes should neither affect their detection nor the detection of the product or other components. Thinner cables are often undetectable or difficult to detect using currently-feasible active depth sensors. In general, cables are characteristic in having (a) a constant colour, (b) a constant, limited width and (c) a length often significantly longer than the width. However, exceptions to these rules (when viewed from one perspective) include ribbon cables and twisted pairs. As cables are commonly found in a large range of colours, a method insensitive to changes in colour is desirable. The current literature presents few cases showing conclusive results for cable detection. Gil et al. [25] employs a Hough Transform-based method that only detects straight sections of wires (see Fig. 4.19). This is sufficient for suggesting some locations for cutting, however detects each wire as many small sections and is incapable of detecting wires that are entirely curved. Hohm et al. [29] suggests scale-space filtering for the detection of cables.

Apart from their flexibility (and hence often-irregular shape), a major challenge in cable detection is that many non-cable objects—e.g. painted lines, light-reflecting edges—share the same distinguishing features of cables. Figure 4.20 shows the results of a detector targeting thin areas located between edges. This is more suitable for the detection of curved sections of wires, however displays an unacceptable rate of false positives due to the aforementioned problems, which are also unsolved by the aforementioned approaches from other literature.

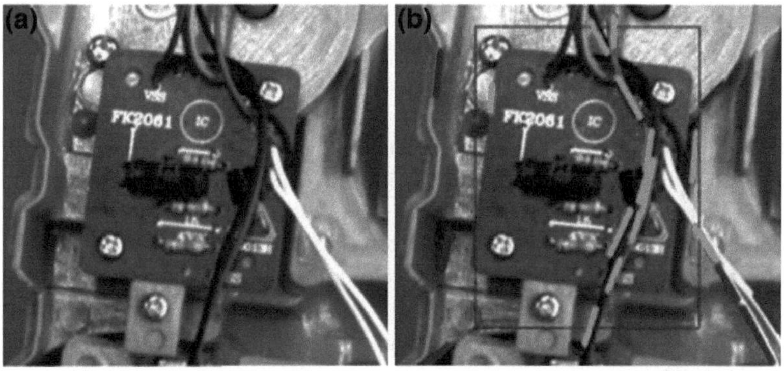

Fig. 4.19 Cable detection with Hough Transform based method [25]. **a** Captured image. **b** Processed image

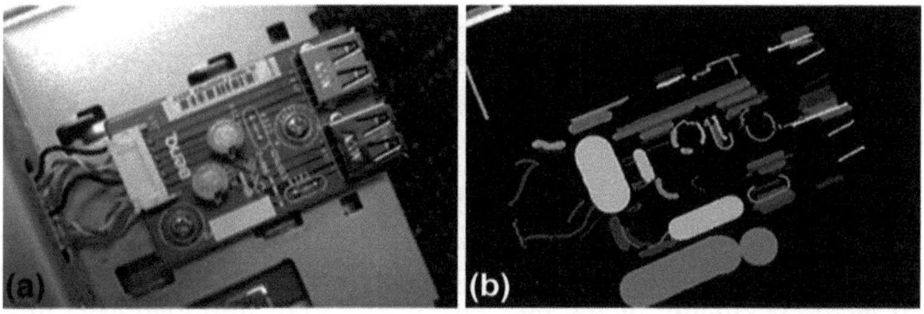

Fig. 4.20 Cable detection targeting thin consistent regions between edges. **a** Captured image. **b** Processed image

Further measures are needed to attain the required robustness for application in industry. Many of the above false positives can be rejected by determining that they lie on a flat surface or form a regular shape. A potential solution also lies in exploiting the knowledge that cables are flexible, by actively testing the flexibility of candidates. Ultimately, the existence or location of wires should be learned with respect to their attachment points, which can also be determined using an active approach.

4.4.4 State Change Detection for Execution Monitoring

A state of disassembly is defined according to the existence of particular main components on the disassembly rig. A state transition occurs when an entire component or significant portion thereof has been detached and significantly moved from its original location. Disassembly state change is a key measure of the success or failure of the current disassembly operation, and is expected to occur when sufficient disassembly operations have been performed. The detection of state change is used for the cognitive function of execution monitoring (further explained in Chap. 5). This book proposes two possible candidate approaches for measuring state change: (a) the *absolute approach* and (b) the *relative approach.*

For the *absolute approach*, the detection of a particular component is repeatedly performed to recheck the properties of the component including its location and existence. State change is indicated if an entirely new set of properties have been detected. This method allows greater flexibility in the case of a more complicated product structure. However, logical ambiguity can arise in case that some parts of a component remain after a significant portion is removed. Remnants may be incorrectly recognised as a new component leading to misperception of the product structure.

For the *relative approach*, the measurement is performed relative to originally-detected properties. The component is detected once at the beginning. Afterwards, incremental change is measured. This approach resolves the ambiguity problem found in the absolute approach. The relative approach is therefore recommended for (semi-) destructive disassembly, where parts of a component may still remain after sufficient operations have been carried out. Only this approach is presented in this book, according to the LCD screen case study.

Detection Algorithm for the Relative Approach

Using the relative approach, state change is detected when a significant amount of change is observed between the current condition and the original condition detected at the beginning of the previous state. Depth information is intuitively more decisive since it represents the physical geometry of the component and is less affected by external factors. However, due to (a) the inaccuracy of the depth sensor, (b) the comparatively insignificant height of some components, and (c) the occurrence of blind areas in the depth data, the colour image is also taken into account to compensate for these limitations. Once a new state is detected, this state is flagged and the new properties stored as a benchmark. Subsequent conditions are compared to this benchmark until the next state change. In the condition definitions, the flagged benchmark will be referred to using the subscript *flag* and the images associated with the current condition with *check.* To disregard the irrelevant surroundings, the difference is only measured within a local *region of interest* (ROI) enclosing the main component.

Equation (4.23) expresses the condition used to determine state change. State change is detected when the measured difference from either the depth or colour images exceeds the thresholds Φ_{depth} and Φ_{colour} respectively. The difference measures are explained as follows.

$$stateChange \equiv \left(diff_{depth} \geq \Phi_{depth}\right) \vee \left(diff_{colour} \geq \Phi_{colour}\right) \tag{4.23}$$

For the *depth image*, one indicator for disassembly state change is the appearance of regions in the second image with height (z) value lower than the corresponding regions on the flagged image [see condition φ_1, Eq. (4.25)]. These regions represent the volume of the component that has been removed. The depth sensing technique is also prone to the occurrence of blind areas due to reflective surfaces (see Active depth sensor in Sect. 4.2.1). Although the depth itself is not sensed in these areas, change in blind area is taken as an additional indicator for state change [see conditions φ_2 and φ_3, Eqs. (4.26) and (4.27)]. A significant change in blind area implies a change in surface to one with a contrasting reflectivity property. The total measured difference with respect to the depth image ($Diff_{depth}$) is the ratio between the number of pixels satisfying at least one of the aforementioned change conditions and the number of pixels within the local ROI [see Eq. (4.24)].

$$Diff_{depth} = \frac{\sum_I 1[\varphi_1 \vee \varphi_2 \vee \varphi_3]}{S_I} \tag{4.24}$$

$$\varphi_1 \equiv \left(z_{i,flag} > z_{i,check}\right) \tag{4.25}$$

$$\varphi_2 \equiv \left(z_{i,flag} \notin \phi_{blind}\right) \wedge \left(z_{i,check} \in \phi_{blind}\right) \tag{4.26}$$

$$\varphi_3 \equiv \left(z_{i,flag} \in \phi_{blind}\right) \wedge \left(z_{i,check} \notin \phi_{blind}\right) \tag{4.27}$$

where φ_i = condition i; I = pixel of the specific ROI; S_I = size of ROI; z_i = height (in direction z_F); and ϕ_{blind} = blind area on the surface of component.

The state difference with respect to the *colour image* ($Diff_{colour}$) is measured using a colour-based histogram comparison in the HSV colour space. The histogram (H_k) is constructed in two channels, H and S, in order to reduce the effects of illumination. $Diff_{colour}$ is obtained by Eqs. (4.28) and (4.29), which are derived from the correlation equation for measuring histogram similarity in [49].

$$Diff_{colour} = 1 - \frac{\sum_I \left(H_{flag}(I) - \bar{H}_{flag}\right)\left(H_{check}(I) - \bar{H}_{check}\right)}{\sqrt{\sum_I \left(H_{flag}(I) - \bar{H}_{flag}\right)^2 \sum_I \left(H_{check}(I) - \bar{H}_{check}\right)^2}} \tag{4.28}$$

$$\bar{H}_k = \frac{1}{N}\sum_I H_k(I) \tag{4.29}$$

where N = number of histogram bins; I = pixel of the local ROI; H_{flag} = histogram of the original condition; and H_{check} = histogram of the current condition.

For the case-study, preliminary tests were conducted by non-destructively removing main components from LCD screen samples. With regards to Eq. (4.23), thresholds of $\Phi_{depth} = 50\ \%$ and $\Phi_{colour} = 75\ \%$ are sufficient to correctly distinguish state change in 95 % of the samples. An example of state change is shown in Figs. 4.21 and 4.22. However, these criteria are subject to slight changes according to the geometry of the components.

Results and Conclusions

The depth criteria are sufficient to identify state change in the majority of the samples. However, due to the limitations of the depth sensor, these criteria fail in the

Fig. 4.21 State change—original condition (flagged images). **a** Colour image. **b** Depth image. **c** 2.5D depth map

Fig. 4.22 State change—the component is removed. **a** Colour image. **b** Depth image. **c** 2.5D depth map

following two circumstances: (a) when removed components have insignificant height, in comparison to sensor resolution or noise; and (b) when a product contains multiple reflective components, with significant portions lying under blind areas. Under these circumstances, an insufficient number of pixels may be counted resulting in false negatives. The colour criterion is robust for state change detection if there is an adequate colour difference between each pair of component types. In the case study, this weaker criterion was found capable of resolving the issues encountered by the depth criteria alone. However, inaccurate assessment was found to occasionally occur during destructive disassembly, as the sensed colour is affected by the presence of dust and fumes. In conclusion, both colour and depth criteria were found to be useful in state change detection, both displaying technical limitations which can be overcome by the other.

4.4.5 *Extensibility*

Only the requirements with respect to the process of disassembly have been discussed thus far. Due to the rapidly-evolving product market, a disassembly system (and corresponding vision system) should not be static in time, but rather, easily extensible to meet future requirements. Software development should occur with the intention of reducing the effort required to implement new functionalities. Modular programming is useful in this context. Additionally, the ability to extend

Table 4.8 Requirements of a GUI with extensibility in mind

Information	Functions	Other requirements
• Raw sensory data • Results from detection • Intended action	• Add a positive example • Flag an error/classify a negative example • Manual control (Advanced: associate a new action to a sensed or detected feature)	• Simple to understand • Simple to operate simple to add new • Sensors/detectors/actions

functionality should preferably remain as open as possible: this is the underlying motivation for a learning-by-demonstration system, and can be applied to the vision system as well. No vision system is guaranteed to perform perfectly, and much time can be saved if *anyone* can make the necessary adjustments, without expert knowledge.

When a database is used for template matching or in training a machine learner, the ability to add new examples to the database, flag errors or add corrections can easily be granted to the operator. For this, the operator also requires access to the information the robot perceives. A *graphical user interface* (GUI) is required, to display the information sensed from the environment and the robot's intentions, as well as allowing user input. Intended actions can be communicated by superimposing action locations onto sensed visual information. This is additionally useful for safety and debugging. Some potential requirements of such a GUI are summarised in Table 4.8.

4.5 Conclusions

The primary advantage of a vision system, with respect to disassembly automation, is in allowing the disassembly system to be robust to the large number of uncertainties present during disassembly. These uncertainties include lot size and product model, as well as potential damage or modification during the usage phase. A vision system allows the system to potentially detect products and variants, components and fasteners, as well as verify the effect of its own actions. This allows the disassembly system to automatically react to sample deviations, unsuccessful actions and potentially unseen models.

However, the implementation of a vision system is a non-trivial task. An overview of sensor technologies was presented, followed by a literature review of common techniques previously used in the field. Those reporting positive results used a clever combination of techniques which take advantage both of the features and requirements of the specific task, as well as relational information with regard to other components or features. Finally, the last section presented an overview of the features and requirements of the primary detection tasks, as well as the requirements for extensibility of such a disassembly vision system.

References

1. Weigl-Seitz A, Hohm K, Seitz M, Tolle H (2006) On strategies and solutions for automated disassembly. Int J Adv Manuf Technol 30(5–6):561–573
2. Scholz-Reiter B, Scharke H, Hucht A (1999) Flexible robot-based disassembly cell. Robot Comput Integr Manuf 15(3):247–255
3. Tonko M, Gengenbach V, Nagel H-H, Schäfer K, Picard S, Horaud R, Mohr R (2009) Towards the integration of object recognition and visual servoing for disassembly of used cars. In: IAR Jahresbericht/Rapport Annuel 1997 (ISSN 0947-0727), Deutsch-Französisches Institut für Automation und Robotik/Institut Franco-Allemand pour les Applications de la Recherche (IAR), IAR Annual Meeting '97, 20–21 November 1997, Gesamthochschule Duisburg/Germany, 1998, pp 79–84
4. ElSayed A, Kongar E, Gupta SM, Sobh T (2012) A robotic-driven disassembly sequence generator for end-of-life electronic products. J Intell Robot Syst 68(1):43–52
5. Büker U, Drüe S, Götze N, Hartmann G, Kalkreuter B, Stemmer R, Trapp R (2001) Vision-based control of an autonomous disassembly station. Robot Auton Syst 35(3):179–189
6. Logitech (2013) Logitech HD Webcam C270. http://www.logitech.com/en-us/product/hd-webcam-c270
7. Fontes YC, Brandão D (2010) Application of stereoscopic vision for disassembly line of electronics devices. In: Proceedings of the 9th IEEE/IAS international conference on industry applications
8. Serranti S, Gargiulo A, Bonifazi G (2011) Characterization of post-consumer polyolefin wastes by hyperspectral imaging. Waste Manage 31(11):2217–2227
9. Freitag H, Huth-Fehre T, Cammann K (2000) Rapid identification of plastics from electronic devices with NIR-spectroscopy. Anal Lett 33(7):1425–1431
10. Imec (2013) Imec first to introduce hyperpectral CMOS camera for snapshot and video. http://www2.imec.be/be_en/press/imec-news/imechyperspectralcamera2013.html
11. PrimeSense (2013) 3D Sensors—PrimeSense. http://www.primesense.com/solutions/3d-sensor/
12. Corporation M (2014) XBox 360—Kinect. www.xbox.com/Kinect
13. Khoshelham K, Elberink SO (2012) Accuracy and resolution of kinect depth data for indoor mapping applications. Sensors 12(2):1437–1454
14. Inc. AC (2014) Xtion PRO LIVE. http://www.asus.com/au/Multimedia/Xtion_PRO_LIVE/
15. Stoyanov T, Mojtahedzadeh R, Andreasson H, Lilienthal AJ (2013) Comparative evaluation of range sensor accuracy for indoor mobile robotics and automated logistics applications. Robot Auton Syst 61(10):1094–1105
16. Foix S, Alenyà G, Torras C (2011) Lock-in time-of-flight (ToF) cameras: a survey. IEEE SENS J 11(9):1917–1926
17. RobotShop (2013) MESA imaging TOF camera SR4500 $44\times35°$ FOV
18. Viggiano JAS (2004) Comparison of the accuracy of different white-balancing options as quantified by their color constancy. In: Proceedings of the SPIE—the international society for optical engineering
19. Bradski G, Kaebler A (2008) Learning OpenCV—computer vision with the OpenCV library. Affine transform, 1st edn. O'Reilly Media Inc., California
20. OpenKinect (2011) Imaging information. http://openkinect.org/wiki/Imaging_Information
21. Siciliano B, Sciavicco L, Villani L, Oriolo G (2009) Vision Ssnsors. Robotics: modelling, planning and control. Springer, London, pp 230–255
22. Vezhnevets V, Sazonov V, Andreeva A (2003) A survey on pixel-based skin color detection techniques. In: Proceedings of the GraphiCon
23. Vongbunyong S (2013) Applications of cognitive robotics in disassembly of products. Ph.D. Dissertation, The University of New South Wales, Sydney, Australia
24. Kovac J, Peer P, Solina F (2003) Human skin colour clustering for face detection. In: Proceedings of the international conference on computer as a tool. The IEEE Region 8
25. Gil P, Pomares J, Puente T, Diaz C, Candelas F, Torres F (2007) Flexible multi-sensorial system for automatic disassembly using cooperative robots. Int J Comput Integr Manuf 20(8):757–772
26. Canny J (1986) A computational approach to edge detection. IEEE Trans Pattern Anal Mach Intell 8(6):679–698. doi:10.1109/TPAMI.1986.4767851

27. Weigl A, Hohm K, Seitz M (1995) Processing sensor images for grasping disassembly objects with a parallel-jaw gripper. In: Proceedings of the TELEMAN telerobotics research conference & ERNET Workshop, Noordwijkerhout, The Netherlands
28. Witkin AP (1983) Scale-space filtering: a new approach to multi-scale description. In: Proceedings of the 8th international joint conference on artificial intelligence
29. Hohm K, Müller Hofstede H, Tolle H (2000) Robot assisted disassembly of electronic devices. In: Proceedings of the 2000 IEEE/RSJ international conference on intelligent robots and systems
30. Rolando Cruz-Ramírez S, Mae Y, Arai T, Takubo T, Ohara K (2011) Vision-based hierarchical recognition for dismantling robot applied to interior renewal of buildings. Comput-Aided Civil and Infrastruct Eng 26(5):336–355
31. Fraundorfer F, Scaramuzza D (2012) Visual odometry Part II: matching, robustness, optimization, and applications. Robotics and automation magazine, IEEE
32. Lowe DG (2004) Distinctive image features from scale-invariant keypoints. Int J Comput Vision 60(2):91–110. doi:10.1023/B:VISI.0000029664.99615.94
33. Bay H, Tuytelaars T, Gool L (2006) SURF: speeded up robust features. In: Leonardis A, Bischof H, Pinz A (eds) Computer vision—ECCV 2006, vol 3951. Lecture Notes in Computer Science. Springer, Heidelberg, pp 404–417. doi:10.1007/11744023_32
34. Agrawal M, Konolige K, Blas M (2008) CenSurE: center surround extremas for realtime feature detection and matching. In: Forsyth D, Torr P, Zisserman A (eds) Computer Vision—ECCV 2008, vol 5305. Lecture Notes in Computer Science. Springer, Heidelberg, pp 102–115. doi:10.1007/978-3-540-88693-8_8
35. Leutenegger S, Chli M, Siegwart (2011) RY BRISK: binary robust invariant scalable keypoints. In: Proceedings of the IEEE international conference on computer vision (ICCV), 6–13 Nov 2011, pp 2548–2555. doi:10.1109/ICCV.2011.6126542
36. Harris CG, Pike JM (1988) 3D positional integration from image sequences. Image Vis Comput 6(2):87–90. doi:10.1016/0262-8856(88)90003-0
37. Rosten E, Drummond T (2006) Machine learning for high-speed corner detection. In: Leonardis A, Bischof H, Pinz A (eds) Computer vision—ECCV 2006, vol 3951. Lecture Notes in Computer Science. Springer, Heidelberg, pp 430–443. doi:10.1007/11744023_34
38. Calonder M, Lepetit V, Strecha C, Fua P (2010) BRIEF: binary robust independent elementary features. In: Daniilidis K, Maragos P, Paragios N (eds) Computer vision—ECCV 2010, vol 6314. Lecture Notes in Computer Science. Springer, Heidelberg, pp 778–792. doi:10.1007/978-3-642-15561-1_56
39. Alahi A, et al. (2012) FREAK: fast retina keypoint. IEEE conference on computer vision and pattern recognition (CPVR), pp 510–517
40. Galleguillos C, Belongie S (2010) Context based object categorization: a critical survey. Comput Vis Image Underst 114(6):712–722
41. Jørgensen TM, Weimar Andersen A, Sloth Christensen S (1996) Neural net based image processing for disassembling TV-sets. In: Solving engineering problems with neural networks
42. Karlsson B, Järrhed J-O (2000) Recycling of electrical motors by automatic disassembly. Measur Sci Technol 11(4):350
43. Viola P, Jones M (2001) Rapid object detection using a boosted cascade of simple features. In: Proceedings of the conference on computer vision and pattern recognition
44. Kopacek P, Kopacek B (2006) Intelligent, flexible disassembly. Int J Adv Manuf Technol 30(5–6):554–560
45. MillerTechnology NEMA specifications. http://www.millertech.com/Technical_Specs.htm. Accessed 18 Nov 2013
46. Vezhnevets V, Sazonov V, Andreeva A (2003) A survey on pixel-based skin color detection techniques. In: Proceedings of the GraphiCon 2003. pp 85–92
47. Chang F, Chen C-J, Lu C-J (2004) A linear-time component-labeling algorithm using contour tracing technique. Comput Vis Image Underst 93(2):206–220
48. Lambert AJD, Gupta SM (2005) Disassembly modeling for assembly, maintenance, reuse and recycling. CRC Press, Boca Raton
49. OpenCV (2014) Histograms. In: The OpenCV reference manual—release 2.4.9.0. pp 295–306

Chapter 5
Cognitive Robotics

Abstract Uncertainties and variations in returned End-of-Life (EOL) products result in complexity at the planning and the operation levels of automated disassembly. These have become critical obstacles in disassembly automation, which lacks the flexibility and robustness of manual disassembly. In this chapter, the principle of cognitive robotics is implemented in disassembly automation, to overcome these problems by emulating the behaviour of human operators. The methodology, framework, and cognitive functionalities with regard to the disassembly domain are explained in this chapter.

5.1 Autonomous Robot and Cognitive Robotics

Autonomous robots are intelligent robots with certain degrees of autonomy that performs tasks by itself with minimal or without human guidance. The tasks are carried out by the Intelligent Agent (IA) which makes decisions according to the information of the dynamic environment sensed. *Cognitive robotics* describes the high-level cognitive functions[1] [1] which allow a robot to reason, revise and perceive change in unpredictable environments. This allows a robot to respond to its environment and complete goals in a robust and adaptive way [1]. Müller [2] illustrated characteristics of classical Artificial Intelligence (AI) and cognitive system in a complexity space which presents a relation between complexity of the tasks and complexity in environment (see Fig. 3.2: Example of serial robots). In this Figure, vertical axis is "flexibility" representing the ability of the agent to deal with complex environment; whereas, the horizontal axis is "specific task success" representing the ability to deal with complex tasks. The classical AI performs complex tasks effectively in non-complex environments, whereas the cognitive system exhibits the opposite characteristics. The cognitive functions allow the

[1] *Cognitive functions (n.)* An intellectual process by which one becomes aware of, perceives, or comprehends ideas. It involves all aspects of perception, thinking, reasoning, and remembering.

S. Vongbunyong and W.H. Chen, *Disassembly Automation*, Sustainable Production, Life Cycle Engineering and Management, DOI 10.1007/978-3-319-15183-0_5

cognitive system to effectively perform simple tasks in a complex environment. Cognitive robotics, used in conjunction with classical AI, therefore allows a system to become more flexible and reliable in dynamic environments [2] (Fig. 5.1).

Cognitive robotics takes the form of various approaches. One practical approach is to use *knowledge representation and reasoning* (KRR) to solve problems in situations where the robot has incomplete knowledge [3]. The knowledge of the environment is represented in a format in which the robot can reason and control its behaviour. The robot interacts with its environment using sensors and effectors. Behaviours are expressed using a high-level programming language based on logic programming. The programs generate action sequences corresponding to conditions: initial states, prerequisites and effects of primitive actions, exogenous events, and the result of sensing. As a result, the robot can interact robustly and autonomously with the external world.

Another important feature distinguishing cognitive robotics from classical AI is the method of interaction with humans: interaction, rather than control, in order to increase the level of intelligence and flexibility of the system [2]. This architecture is based on *a closed perception-action loop* as shown in Fig. 5.2 [4]. The behaviour of the robot is organised by three main elements: (a) *learning and reasoning*,

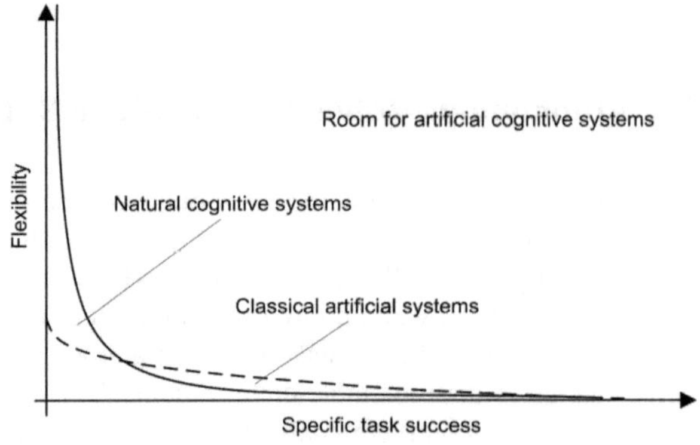

Fig. 5.1 Complexity space for intelligent agent [2]

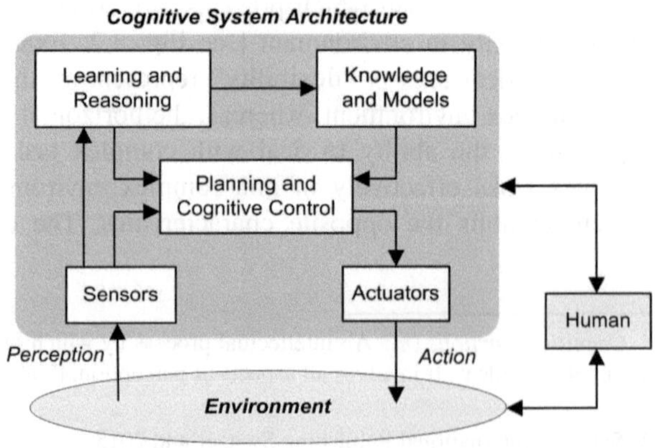

Fig. 5.2 Cognitive system architecture with closed perception-action loop. Adapted from [4]

(b) *planning and cognitive control*, and (c) the *knowledge model*. Interaction with the external world occurs via the sensors and actuators, which are associated with perception and action respectively. The human is considered part of the external world and can interact with both the environment and the robot. This architecture was originally implemented as part of "*Cognitive Factory*" in which human operators work side-by-side with automation that is flexible, reliable, and safe.

Cognitive robotics has recently been presented in much research in many fields. The Cognitive Factory is one of the state-of-the-art applications of cognitive robotics in industry. This project is developed by CoTeSys (Cognition for Technical Systems), a company that applies cognitive robotics functions to industrial activities, especially production [5]. The project integrates the key activities of production: (a) system monitoring and planning; (b) material and parts condition; (c) work piece assembly; and (d) human-machine co-operation. Four main functions of cognitive robotics are implemented: *perception*, *learning*, *knowledge*, and *planning*. These functions allow the robots to monitor the product and carry out the assembly process with respect to prior knowledge of the assembly plan. A learning module makes the system self-optimise and derives a better assembly sequence from past incidents. Consequently, an optimal sequence plan can be autonomously found by the planning module. With respect to skill acquisition, self-adaptation, and self-modelling, the cognitive factory is flexible enough to operate with a variety of products. In comparison to classical manufacturing, the Cognitive Factory is hence capable of achieving higher productivity and flexibility in the production line, as illustrated in Fig. 5.3 [6].

In the field of disassembly, much research regarding automatic disassembly cells has been conducted using classical AI. However, the flexibility of the existing

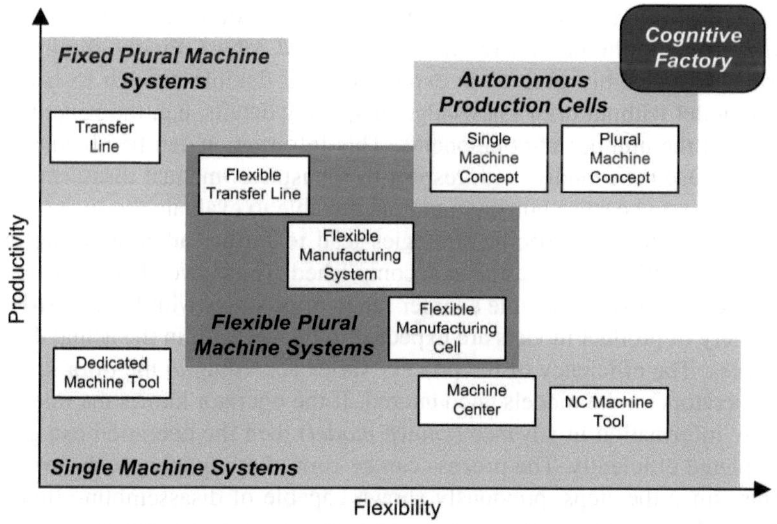

Fig. 5.3 Classification of various types of manufacturing with respect to productivity and flexibility [6]

system is very limited in dealing with variations in the EOL product in the planning and operational levels. Due to the aforementioned advantages of cognitive robotics, it appears beneficial to employ these concepts in handling the uncertainties in disassembly domain. The disassembly process can be considered a dynamic external world with which the cognitive robotics system interacts. This chapter presents the application of cognitive robotics in product disassembly.

5.2 Concept Overview

In comparison to automated systems, human operators are far more capable of dealing with uncertainties in the EOL products returned. Therefore, the principle of cognitive robotics has been used to emulate the behaviour expressed by a human operator during the disassembly process. These behaviours are expected to address the uncertainties in product and process that have thus far hindered the industrial application of robotic disassembly. Common behaviours of the human operator and their implementation using cognitive robotics principles are explained in this section.

5.2.1 Human Driven Disassembly

Human operators are expected to intuitively overcome the aforementioned uncertainties in disassembly. Decisions are taken based on prior expertise, by adapting past experiences to the perception of the currently-encountered situation. The process performed by human operators is *flexible* and *robust*. Firstly, the intuitively-performed disassembly process is expected to be flexible enough to handle any product model without prior knowledge of specific details, e.g. the structure of the product and the number of components. This information can be obtained during the process itself. Secondly, with respect to robustness, manual disassembly has a high success rate because human operators are able to evaluate the success of each step. Hence, failures in specific strategies lead to further attempts using alternative methods until the task has been accomplished. These two characteristics allow human operators to disassemble a larger range of products with fewer restrictions.

A variety of product models are expected to be met with in the actual disassembly process. The efficiency of the process varies according to the prior knowledge of the operators on the models encountered. If the operator knows the relevant disassembly information in advance (*known model*) then the operation can generally be completed efficiently. The process can be carried out quickly with few fruitless attempts, since the steps, previously shown capable of disassembling this model, are already known. Minor physical uncertainties are expected but can generally be compensated intuitively.

On the other hand, difficulties may arise in the case that the model is new to the operator (*unknown model*). Since the specific information is unavailable, the operator attempts previously-known strategies to discover a suitable method for the new model. This process may be carried out awkwardly, state by state using trial and error, for the first few attempts to disassemble each new model. In each disassembly state, the operators may spend more time to locate or identify main components. A number of unsuccessful attempts may be made before the components are successfully removed and the operator can proceed to the next state.

Possible operations can be logically selected based on the operator's perception and past experience. Concurrently, during the process, the operator gathers information: product structure, component-related details, disassembly operations and relevant parameters. The operator learns an appropriate process for the new model by considering the relation between actions and their consequences: success or failure. The knowledge base (KB) is continuously built up and recalled when needed. This process is repeated in all disassembly states and finished when the goal state has been reached. After a certain number of samples have been successfully disassembled, the operator should have enough experience to effectively disassemble this particular model when it is again encountered. This human behaviour in disassembly process is illustrated in Fig. 5.4.

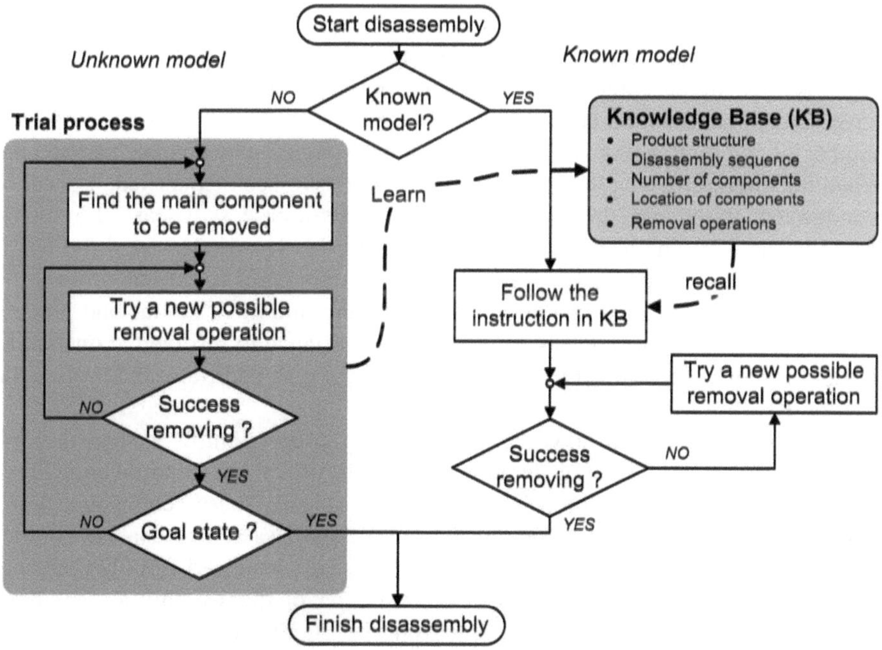

Fig. 5.4 Behaviour of human operators in the disassembly process

According to this behaviour, the characteristics of the human operator that influence the flexibility and the robustness of disassembly are summarised as follows:

- Ability to perceive product structure during the disassembly process;
- Ability to assess the outcome of each performed operation and perform alternative operations if the first attempt fails;
- Flexible operation schemes that can be adapted to other physically similar components; and
- Ability to learn from past experience and adapt to previously unseen cases.

These characteristics are emulated by using the principle of cognitive robotics which is explained in the following sections.

5.2.2 Agent Emulating Human Behaviour

The aforementioned human-like behaviour needs to be adapted to be suitable with the *closed perception action loop architecture* and capabilities of the automation. In considering the limitations of automation, assistance is introduced to the behaviour control in order to help the robot to handle unresolved situations. The cognitive robotic agent (CRA) first determines whether or not the model of each sample is already known in the existing knowledge base (KB) . If the model is being seen for the first time (unknown), the agent[2] goes through the trial process in which actions are executed based on general operation plans according to the perceived components in the current state. The removal process is carried out with a number of attempts using different operation strategies and process parameters. When the alternatives available to the automation have been exhausted, assistance is requested. In the end, the model-specific successful process is learned by being stored in the KB. The agent is now able to recall this knowledge from the KB if this model of product is seen again (known). In this case, the agent follows the instructions in the KB on default. In case of failure due to variations and process uncertainties, the agent requests for additional assistance to resolve the uncertainties and achieve the goal. The agent also learns the new additions and revises the existing KB (Fig. 5.5).

Regarding the process flow, the agent drives the disassembly system through the states of disassembly by generating a sequence of actions using its knowledge of the external world and the behaviours at its disposal. These actions are physically performed by the robot. The transition from one disassembly state to another occurs when a main component has been successfully removed from the original location, and occurs after a sufficient combination of disassembly actions have been performed. This process continues from the initial state of the product until the goal state has been reached.

---

[2] In this chapter, "*agent*" refers to "cognitive robotic agent (CRA)".

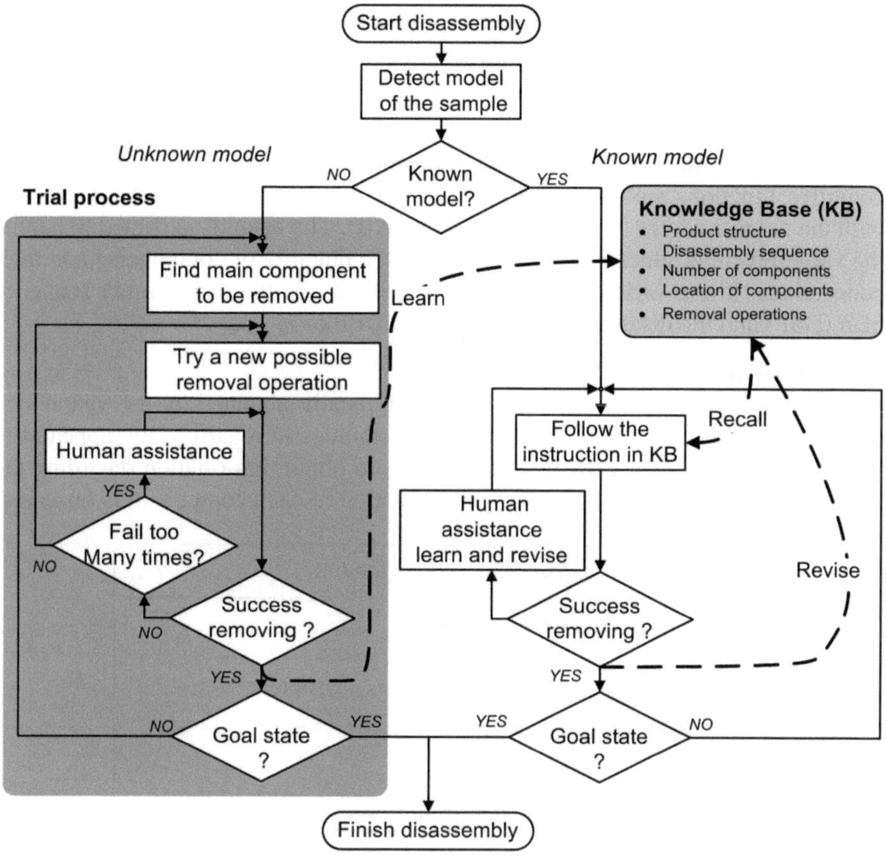

Fig. 5.5 Human behaviour emulated by cognitive robotic agent

In perspective of the information available to the agent, the physical disassembly operations have a non-deterministic effect on the world due to uncertainties in the process and product condition. Therefore, this automated process is considered an *open-world execution* since the incomplete knowledge of the world needs to be sensed on-line during the process. The interaction with the external world is described in the next section.

5.3 System Architecture and Cognitive Functions

5.3.1 Operation Modules and Uncertainties

The uncertainties in the process are typically caused by the variations in the products to be disassembled. The types of uncertainties are expected to be effectively handled by the appropriate modules of the system. The connection between the modules and the uncertainties to be addressed are illustrated in Fig. 5.6 and summarised in Table 5.1. The modules are explained as follows.

Cognitive robotic module (CRM)
This module is an AI agent that controls the behaviours of the system according to the cognitive functions and relevant knowledge of the product and process. The cognitive functions are expressed by the cognitive robotic agent, which interacts with the other modules and knowledge base (KB). The agent is expected to handle the variations in product structure relating to the quantity of components and their connections. These variations result in uncertainties in the disassembly sequence plan (DSP) and the disassembly process plan (DPP).

Vision system module (VSM)
Information regarding the physical world is typically obtained by the vision system. This module is designed to handle uncertainties in regard to the appearance variations of components. Its main function is to obtain information regarding the underlying components. The quantity and location of each component is observed.

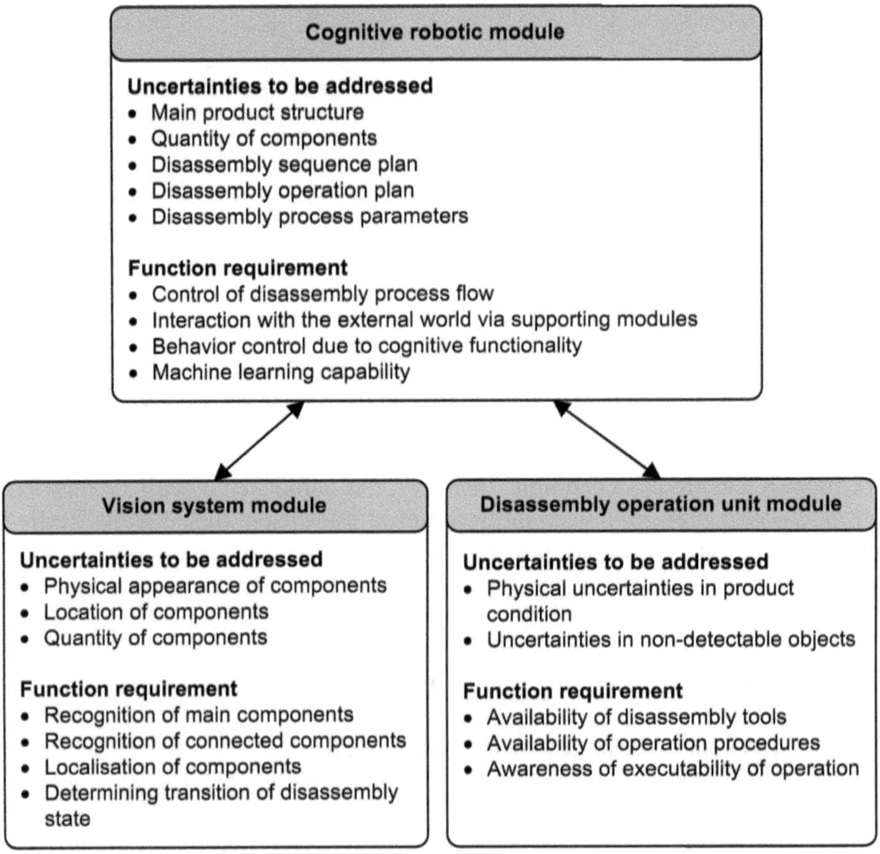

Fig. 5.6 Specification summary of the robotic disassembly system

Table 5.1 Uncertainties in disassembly process

Uncertainties	Specific issue of uncertainties	Module		
		CRM	VSM	DOM
EOL condition	Physical product conditions			•
Variety in the supplied products	Main product structure	•		
	Physical appearance in the components		•	
	Quantity of the component	•	•	
	Location of the components		•	
Complexity in process planning and operations	Disassembly sequence plan	•		
	Disassembly operation plan	•		
	Disassembly process parameters	•		
	Uncertainties in the non-detectable objects			•

In addition, visual input is used in the execution monitoring process in order to determine the success or failure of physical operations.

Disassembly operation unit module (DOM)
The operation units consist of actuators and sensors having direct contact to the product. The main functions are to detach the connections between the main components and remove these components from the remaining parts of the product. Robot arms equipped with interchangeable disassembly tools and grippers are generally used for these purposes. Force and tactile sensors are used for enhancing the system's capability to perceive visually unobservable information important to the disassembly process. The destructive disassembly approach is less sensitive to variations in the EOL product condition and reduces the complexity in detaching connective components. However, the desired outcome of EOL treatment must be taken into account.

5.3.2 Architecture

The architecture of this system is developed based on the *closed perception action loop* architecture [4] in which the CRM interacts with the physical world via the disassembly cell and human input. The system architecture is shown in a diagram in Fig. 5.7. The CRM connects to the physical world through sensors and actuators which are the VSM and the DOM in the disassembly cell. The behaviour of the system is expressed though these modules. In case that the system cannot resolve the problems autonomously, human experts are able to assist by demonstrating the operations or revising the original belief in the product and process. This framework was earlier presented in [7].

The CRM consists of the CRA and KB, which interact internally with each other. The CRA plays an important role in controlling the behaviour of the system.

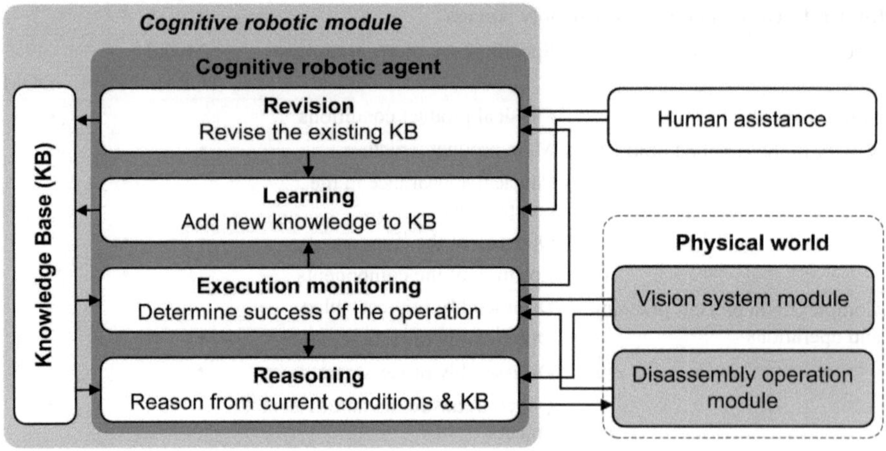

Fig. 5.7 System architecture in cognitive robotics perspective

This behaviour is expressed according to four cognitive functions: (a) *reasoning*, (b) *execution monitoring*, (c) *learning*, and (d) *revision*. These core functions allow the disassembly system to react to the dynamic external world in a flexible and robust way. The CRA typically interacts with the KB by consulting or modifying specific knowledge according to the product being disassembled. The CRA consists of generic procedures and rules, which are used for the general disassembly process. Meanwhile, the KB contains information that is more specific to the product models. The cognitive functions in CRM and the KB are explained as follows.

Reasoning

The disassembly process can be represented as *choice-points* through which the CRA drives the system until the goal is achieved. Reasoning is used to prune these choice-points according to the predefined rules, the disassembly state perceived by the sensors, operation success or failure as measured by execution monitoring, and the existing knowledge in the KB. As a result, the CRA can react logically to the external world via the DOM.

Execution monitoring

This function measures the accomplishment (success or failure) of the removal of components at the planning and operation levels according to the predefined rules. At the planning level, information regarding the change of disassembly state is supplied by the VSM. This information can be used to determine the main structure of the product as disassembly progresses. At the operation level, the outcomes returned by the actuators performing the assigned tasks further allow the reasoning function to perform parameter adjustments where necessary. Subsequently, the corresponding operations and plans are learned and revised.

Learning
This function typically deals with the *model-specific knowledge* which is information specific to a particular model of product. This knowledge is obtained during the disassembly process. Significant information contributing to a successful component removal is stored in the KB. Consequently, the CRA can use this knowledge in subsequent processes for the disassembly of this already-known model. This knowledge can be obtained from two sources, the reasoning process and demonstration by assistance.

Revision
This function is used to revise the existing knowledge in the KB. The aim is to improve the efficiency of the process of disassembling new samples of the previously seen model. Some improvement is expected to result from removing redundant operations from the previous process.

In addition, assistance may be used to correct a false original belief causing failures in component removal. The system autonomously builds its belief according to sensor perception. Since the sensors are not perfect, a certain degree of error is expected which results in imprecise belief regarding the product and operations.

In a more advanced approach, the CRA may also have the ability to modify its own program, revising its general (*non-model-specific*) behaviour according to the new knowledge received. However, it should be noted that this self-modification ability has not yet been implemented in disassembly automation.

Knowledge base
The KB contains knowledge associated with the DPP for both the general and the model-specific disassembly processes. The general knowledge is used in the case of disassembling an unknown model. This knowledge consists of constraints and parameters for using with the generic rules are stored. Conversely, the model-specific knowledge is applied when disassembling a known model. This knowledge consists of (a) *disassembly sequence plan* (*DSP*), (b) *disassembly operation plans*, and (c) *process parameters*. The KB will be continuously expanded in regard to learning and revision as disassembly is carried out.

5.3.3 Language Framework and Interactions

In this book, the CRA is modelled using the cognitive robotics framework based on *Situation Calculus* [8]. *IndiGolog* is the programming language selected to model the system. The main advantage of this language is to support on-line execution. The language supports sensing and exogenous actions which allows the agent to deal with incomplete knowledge perceived from the external world [9, 10]. IndiGolog can be implemented on various language platforms; in this case, *Prolog* is selected [11]. Prolog is a logical and declarative language widely used in AI. Prolog also has an inference engine which is used to find the solutions from the given rules. This characteristic allows the researchers to develop the system that is capable of dealing with complex problems with less and simpler code needed in comparison to imperative language. The program is more capable of dealing with the complex problem more According to the characteristics of the high-level programming language IndiGolog, the CRA and the states of the world can be modelled using *behaviour specifications* and *domain specifications*.

First, the *behaviour specification* is defined as procedures for performing complex actions in order to achieve the desired goal. The control structure uses procedural language control structures (conditional statements, testing, loop statements, action sequences, and sub-procedures) together with nondeterministic constructs for planning and search (nondeterministic choices of argument, concurrency, interrupt, and searching). All of these components determine the behaviour of the agent and provide mechanisms to prune the search space of the nondeterministic choices of actions through the disassembly domain.

Second, the *domain specification* represents the history of the world as a first order term called a *situation* which is a sequence of actions executed inside and outside the CRA. The properties of the world are described by predicates called *fluents*. The change of the fluents is characterised by *successor state axioms* which describe the relation between fluents, actions, and preconditions. The value of the fluents is dynamically changed after the corresponding actions have been executed when the defined preconditions hold. The successor state axiom of fluent f can be written in the form of $f\left(\vec{x}, do(a, S)\right)$ where the free variables $\vec{x}$ will be changed by executing an action a in the situation S in which the preconditions hold. Actions can only be executed when their preconditions hold. Actions are categorised into three types:

- Primitive actions,
- Sensing actions, and
- Exogenous actions.

The *primitive actions* are basic actions used for performing a task, in which the purpose can vary from internally modifying system variables to physical operations that affect the external world. *Sensing actions* are a type of action executed in order to obtain information, particularly from the external world through sensors. *Exogenous actions* are actions, which are sent from the external world.

All of these components are used to systematically describe the behaviour of the system and the world. The CRM is modelled by both the domain and the behaviour specifications. The primitive actions are defined and used according to the behaviour specification. The fluents are updated according to performed actions and sensed information. Consequently, the CRM performs tasks according to the desired behaviour with respect to the sensed condition of the world.

As shown in Fig. 5.8, communication between the modules occur using abstract information in the form of actions and fluents.

Sensing actions are sent to the corresponding modules as the result of the internal process needing more information from the external world. Subsequently, the sensing result is sent back to the CRM in the form of fluents. The data structure of the fluents varies according to module and information conveyed. For example,

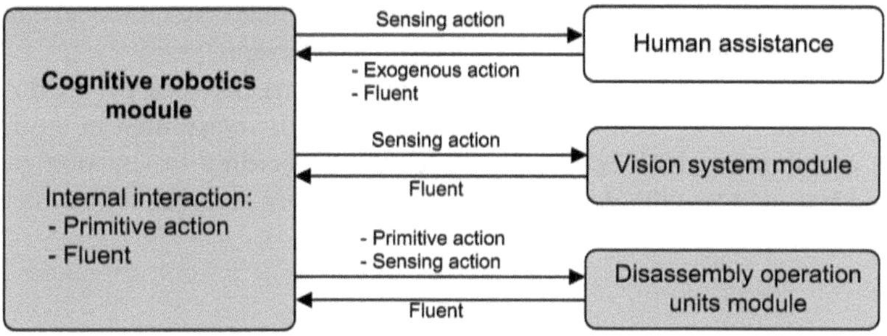

Fig. 5.8 Interaction with actions and fluents

the location of a component obtained by VSM may be expressed using *box*(x_1, x_2, y_1, y_2, z_1, z_2). *Primitive actions* are generally sent to the DOM to request physical operations, e.g. cutting at a particular position *cut(x, y, z)*. Fluents are normally sent back to the CRM after the execution has been finished, and inform the CRM about the success or failure of the primitive actions. Thirdly, *exogenous actions* are sent to the CRM in the case of a human demonstration. A demonstration can also be in the form of a fluent.

The CRA only proceeds once the result fluent is received from the respective module, and continues in this fashion until the desired task is complete. This operation routine is shown in Fig. 5.9. In general, this control structure is sufficient to conduct complex processes such as automated disassembly. However, the control

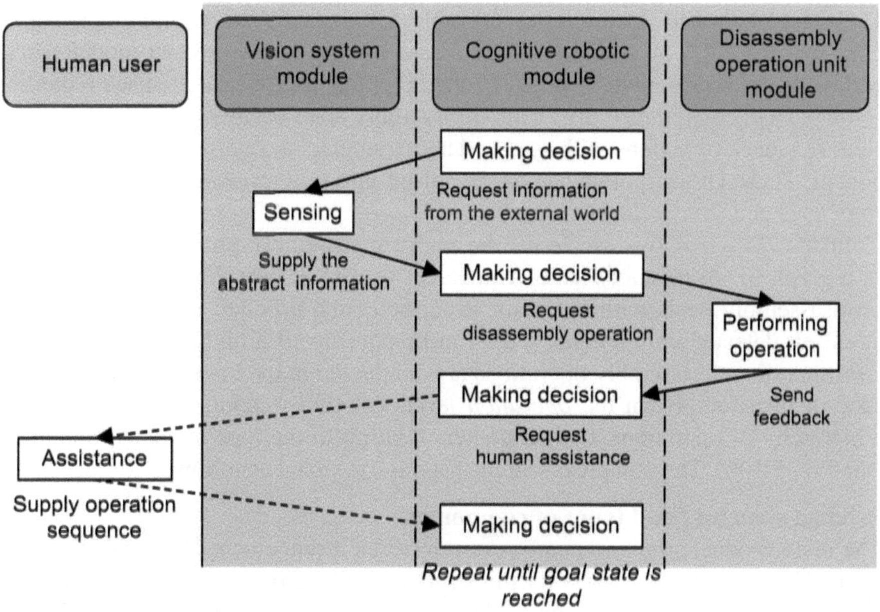

Fig. 5.9 An overview of the common operation routine

structure can be modified to be more efficient and complex by taking advantage of other features of IndiGolog, e.g. concurrency and interrupts.

In conclusion, a general framework of interactions between the modules has been presented in this section. Abstract information is transmitted in the form of actions and fluents which can be clearly defined according to situation calculus. It should be noted that this language framework is only a guideline in which other languages can be considered.

5.4 Basic Behaviour Control

The main function of the CRA is to prune the search space according to the cognitive functions of reasoning and execution monitoring. The search space describes the domain of disassembly processes consisting of variations of plans and operations. For a new product model, the information regarding the product and process is missing. Consequently, the search space has been limited to general strategies broad enough to apply to various models of the product. Specific operations, not encompassed by these general strategies, are further acquired as needed from human demonstration.

5.4.1 Disassembly Domain

The aforementioned uncertainties in Table 5.1 describe the variations in conditions that the CRA must handle while disassembling a new product model. These variations are represented as possible choice-points in the search-space regarding action sequences. Transitions in the disassembly state occur by executing a sufficient sequence of actions with appropriate parameters to remove a particular component. These choice-points can be organised into two levels: (a) the *disassembly state level* and (b) *operational level.*

In the *disassembly state level*, the entire disassembly process is represented as a graph of disassembly states equivalent to a DSP (see Fig. 5.10a). The state of disassembly deals with the main structure of the product; decisions are made based on detected components. This can be considered a high level planner. The corresponding disassembly operations are conducted in the lower level within each state of disassembly. In the *operation level*, the choice-points are formulated by a hierarchy of operations and parameters relating to the type of main component (see Fig. 5.10b). The complete search space is explained as follows.

Product structure and types of component

The main product structure can be considered an arrangement of the main and the connective components. Variations in structure may be presented in different models within the same product family. The state of disassembly is defined here as the set of main components detected at a particular time.

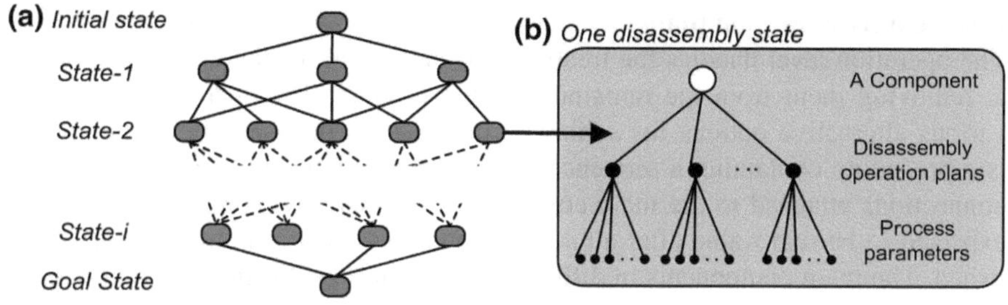

Fig. 5.10 Choice-points in disassembly domain. **a** Disassembly states, **b** operational level

The complete main product structure can be obtained by tracking the state of disassembly. The vision system observes the components belonging to each state of disassembly to obtain the following four categories of abstract information:

- Types of main component;
- Quantity of the main components;
- Types of connective components; and,
- Quantity of the connective components.

For example, the structure in the minimal form of a liaison diagram in Fig. 5.11b can be obtained from the detection results in Fig. 5.11a. This liaison diagram shows the connections between two pairs of main components. PCB1 and carrier are connected by 6 screws and PCB2-carrier connected by 5 screws. The accuracy of the reconstructed product structure depends heavily on the accuracy of the vision system in recognising the components.

Therefore, the specific product structure does not need to be known a priori. This flexibility reduces the need for a pre-defined product structure which is rarely known in advance by the EOL product treatment facility. Finally, the CRA executes suitable plans according to this visual input and the reasoning process explained in the following section.

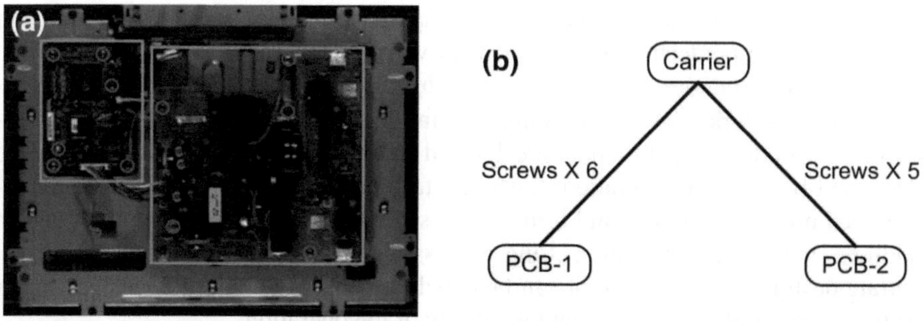

Fig. 5.11 Representation of a product structure in a disassembly state. **a** Detected components in a disassembly state, **b** minimal form of liasian diagram

Disassembly operation plans

The operation level handles the treatment of individual main components, aiming at removing them from the remainder of the product. The operation plans offer various alternative options for achieving this task. A disassembly operation plan is a procedure containing a sequence of primitive actions used to disestablish the connections attached to the main component. The underlying main component is expected to be removable after all corresponding connections have been disestablished. The main components in different models are fixed using different connective components at different locations, some of which are not visually detectable. Hence, multiple operation plans are developed for each type of main component, associated with varying levels of component damage and success rate of the removal process.

A general form of the operation plan is in Procedure (5.1) which represents one of the plan variations (plan-k) for the component c_l with a set of parameters t. During the disassembly process, the agent attempts to execute the operation plans one by one. Alternative sets of operation plans can be executed in different conditions, including type of components and success of previous operation plans. As part of trial-and-error strategy, the agent modifies the operation plan until the main component is removed (i.e. until a state change).

```
Proc operationPlan(c_i, k, t⃗)        // plan k with parameters t⃗ for component c_i
  primAction_1(t_1);                  // execute primitive action with t_i ∈ t⃗
  primAction_2(t_2);
  ...                                                                      (5.1)
  primAction_n(t_n);
endProc
```

Process parameters

The variation of process parameters also forms choice-points, generated during the process within the range limited by predefined constraints. The process parameters (e.g. cutting depth) directly relate to primitive actions where the low-level operations are conducted. In order to remove a component, the agent executes primitive actions within the operation plan, varying the value of each related parameter until the component is removed. Such a trial-and-error approach is necessary if the parameter value cannot be predetermined. The critical value is generally unknown due to two sources of uncertainties: (a) *unobservable objects* and (b) *imprecision of the operation*. One clearly demonstrative example is in the destructive disassembly of a plastic cover. Hidden snap fasteners are located underneath the cover and need to be destroyed by a cutting operation. Since the fastener locations can neither be predetermined nor detected, trial-and-error is necessary to find the critical depth of cut. In addition, tool wear during the cutting process places uncertainties around the exact cutting destination. Therefore, it can be concluded that the ability to vary parameter values increases the robustness and flexibility of the operation.

Predefined constraints to parameter values include the deepest allowed cutting level, maximum offset from the component border, maximum feed and cutting speed. Parameters can typically be varied over a continuous range. To reduce the number of possible choice-points, a discrete set of possible choices are provided. Equation (5.2) shows such choices expressed as a simple set. With respect to the logical reasoning of the agent, the choices can also be represented as a set of preconditions (initial and final value) and rules (strategy to step changing) [e.g. Eq. (5.3)].

$$process\,parameters = \{a_1, a_2, a_3, \ldots a_n\} \tag{5.2}$$

$$z_{min} \leq z_{cut} \leq z_{max} \text{ and } z_{cut} = z_{old} + \Delta z \tag{5.3}$$

5.4.2 Reasoning

In the field of artificial intelligence, reasoning is one of the main functions used in automated planning. The intelligent agent autonomously carries out the process from the initial condition to the goal state under the given constraints, which model the agent and the external world. A number of reasoning techniques are widely used [12]. This book presents only the simple method of *rule-based reasoning*. The interaction with the information from the external world is also emphasised. More complex reasoning methods can be applied using this proposed framework.

According to the cognitive robotics architecture presented in Sect. 5.3, the agent prunes the search space by reasoning about the conditions at each state of disassembly. These conditions occur in four categories:

- The disassembly domain representing plans and operations;
- Condition of the current disassembly state as perceived by the sensors;
- Success or failure of removal operations; and
- Existing knowledge regarding the process for specific models of products.

The disassembly domain shown in Fig. 5.10 should be reconsidered due to these perspectives. As a result, the more specific version is shown in Fig. 5.12 where one state of disassembly is emphasised.

For an *unknown* model of a product, the agent performs disassembly based on trial-and-error throughout the disassembly domain. At the planning level, the agent reasons about the main structure and the main components detected in each state. Relations between the combination of components and appropriate treatment plans must be defined. In this case, a rule-based heuristic is used. A set of rules are predefined by a human expert according to each product family as a broad guideline for the agent. To determine the correct level of specificity for the rules, one must consider the trade-off between the achievement rate and flexibility of the system. Highly specific rules may efficiently handle specific product models but are then less capable of disassembling unknown models. The proposed set of rules only takes the existence (presence or absence) of the main components into account. The existence of components is a priori incomplete knowledge that is obtainable

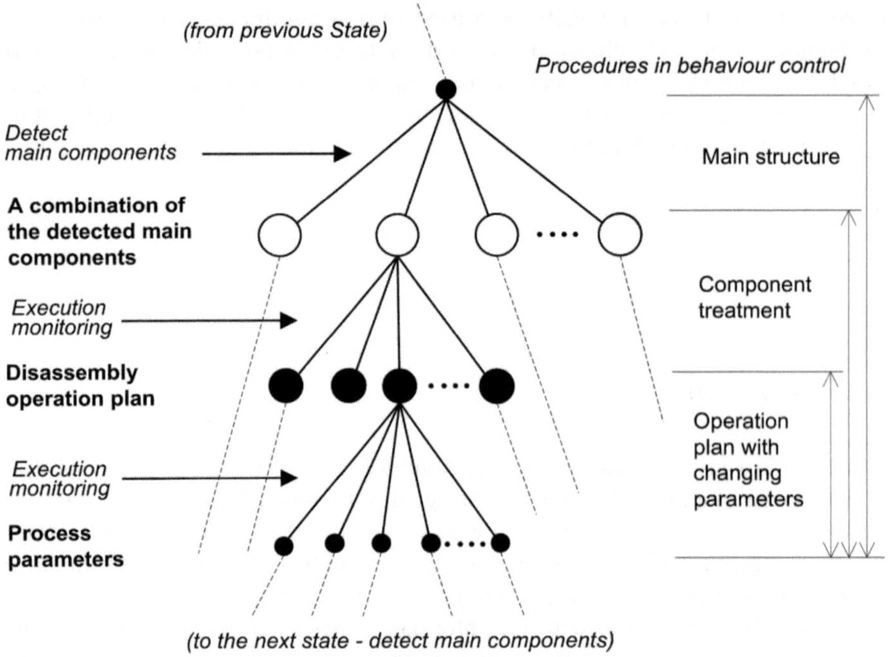

Fig. 5.12 Behaviour control in regard to the disassembly domain

via the sensors. The possibilities, regarding components that can be detected in each state, are also modelled.

In Statement (5.4), the component-k (c_k) is treated if the left hand side of $rule_k$ is satisfied, i.e. if the components c_i to c_n are detected and c_j to c_m are absent. A complete set of rules is required for each component to be treated. An example is given in the case-study in Chap. 6.

$$rule_k : (c_i \wedge c_{i+1} \wedge \ldots \wedge c_n) \wedge \ldots \wedge \left(\neg c_j \wedge \neg c_{j+1} \ldots \wedge \neg c_m\right) \rightarrow treatment(c_k) \quad (5.4)$$

where $c_k \in \{c_i, \ldots c_n, \ldots c_j, \ldots c_m\}$.

After the rules are established, the CRA is able to reason about the corresponding disassembly operation plans and their process parameters respectively. The CRA prunes the process parameter options belonging to a disassembly operation plan until either the goal is achieved or all available options have been exhausted. The goal in each disassembly state is the removal of a particular main component. In the case that the component is not removed, the agent backtracks to the next available operation plan and prunes the corresponding process parameters on the new operation plan. This pruning can be considered a depth-first search. The *execution monitoring* process conducted after physical operation determines whether or not the component has been removed. After achieving the goal, the CRA returns to the planning level to decide the main component to be treated in the next state.

This process continues until reaching the ultimate goal, in which the entire product has been disassembled. This component treatment process can be written as Procedure (5.5).

```
Proc treatComponent(c_i)                    // treat component c_i
  while(¬stateChanged)                      // repeat operaiton if state not changed
    operationPlan(c_i, k, t̄)                // execute plan k with parameters t̄
    removeComponent                         // try removing the component
    detectStateChange                       // execution monitoring - success/failure
    if (¬stateChanged)                      // if fail to remove component
       then vary(k); vary(t̄)                // change plan and parameters
    endIf
  endWhile
endProc
```
(5.5)

For a *known* model, the agent performs the process according to the knowledge in the KB, built up from previous operations. The process is performed in a straightforward manner from one state to another with minimal trial-and-error. However, the sensing of components and execution monitoring are still necessary to maintain the robustness and flexibility of the system. The agent needs to be capable of adapting to possible changes caused by process uncertainties and variations in the product that remain unknown until a certain disassembly state is reached.

A challenge occurs in reasoning regarding destructive disassembly, because it is an irreversible process. When only dealing with reversible processes, as may be the case in non-destructive disassembly, the agent can backtrack to other options in a straightforward manner. In this case, if a particular operation fails, the agent can return the external world to the same condition as before the failed operation was attempted. Then the next operation will attempt subsequently. For the destructive operations, permanent damage on the components can introduce new uncertainties resulting in more complex conditions for backtracking. These uncertainties can only be addressed by using additional strategies, e.g. statistical reasoning approaches, a physical model or sensing techniques for detecting damage, operation plans capable of handling a wider range of uncertainties. This issue will be explained in the case-study in Chap. 7.

5.4.3 Execution Monitoring

Execution monitoring is considered an important function of closed-loop control providing the feedback of the process and operation. In this case, the execution monitoring is performed at the planning and the operational levels where the outcome is either success or failure. The motion control in motor-sensor level is not included in this scope.

Execution monitoring at the planning level deals with the main structure in the disassembly domain (see Fig. 5.12). In each state of disassembly, it determines whether the current main component has been detached within the component treatment according to the reasoning process. As a result, the agent will proceed to the next state if the removal is successful. Otherwise, the trial-and-error process using other available options of the disassembly plans and process parameters will be continued. A general execution monitoring is shown in Procedure (5.5) where *detectStateChange* is executed. Vision system is generally used for assessing the removal process by detecting the change of the components. Our proposed technique (see Sect. 4.4.4) indicates that the process succeed if the volume and appearance are substantially changed from the beginning of the state.

At the operational level, the execution monitoring focuses on executability of the operation in regard to the physical constraints. Accessibility of the desired component has become difficult problems in complex products. In order to approach the object in exact position and orientation, complete knowledge of the model of the disassembly tool and the working environment, i.e. remaining of the product, are needed. This knowledge can be obtained by predefined geometrical models of the product and the tools. Actual information at the time performing the operation can be perceived from the sensors also. However, inaccuracy of this information potentially occurs due to the physical uncertainties in both cases. Therefore, the feedback indicating the successful approach is used to address this inaccuracy. Since this uncertainty involves the physical condition which is visually non-detectable, force and tactile sensors are used to perceive the information regarding these constraints. Subsequently, the agent will be able to try executing the current operation plan with the alternative available process parameters. If the execution can be completed without any crash, the next operation in planning level will be considered.

The execution monitoring is implemented in both unknown and known models cases. It is crucial in the unknown model case since less information of the process is known a priori. During the trial-and-error process, regarding the learning process, the KB will be built from critical information according to the executed plans, the operations, and the process parameters. The critical information is collected when the state has changed. In case of known model, the agent has the similar behaviour but execute less frequently since most knowledge of a particular model of product supposes to be already built from the first time seeing that model. Therefore, the learning happens in the form of revision. The learning and revision are explained in the following section: Advanced behaviour control.

5.5 Advanced Behaviour Control

The advanced behaviour involves the learning and revision process of the CRA [13]. The main objective is to collect significant pieces of information extracted in the current disassembly process and reproduce the same disassembly outcome more efficiently. According to the basic behaviour, the advanced behaviours are expressed in every level of plan and operation execution during the disassembly process. As the learning process, the knowledge is initially obtained in the process for the unknown model. Later on in the revision process, when the known model has been recognised, this existing knowledge is utilised and the agent continuously obtains more knowledge to revise the existing KB for increasing the process performance. The KB is revised repeatedly every time the known model has been disassembled. These activities are conducted autonomously during the process. In addition, human operators involve certain complex situations where the CRA cannot carry out by itself.

In this section, the knowledge base will be explained first followed by the two cognitive functions: learning and revision. Assistance will be explained as parts of learning and revision.

5.5.1 Knowledge Base

To achieve an efficient disassembly, information regarding product and process (e.g. the specification of the main components, their connective components, and detachment method) is expected to be accurately determined from the previous disassembly process. The agent will disassemble according to this information; therefore, the time consumption and redundant operations due to further trial-and-error will be minimised. The knowledge in the KB representing a successful sequence of choice-points in the disassembly domain is used.

The location of components and their corresponding tool executing position are ones of the important process parameters. To accurately determine the specific location, currently, the proposed methods for learning and revision are developed for modifying the *model-specific knowledge* rather than the *generalised rules*. The model-specific knowledge is used since the exceptional features and uncertainties vary from a model to another model. These locations are arbitrary depending on the design of each model and there are other relations presented. In addition, high success rate of disassembly can be achieved by using the model-specific knowledge since it is directly obtained from successful disassembly cases. The improvement of the process is expected to be seen within first few revisions due to the promising success rate.

In the KB file, the knowledge is stored in the form of structured data according to each specific product model. In this book, the knowledge is presented in the form of Prolog facts due to compatibility with IndiGolog as mentioned in Sect. 5.3.3. The inference engine in Prolog is used to match the input queries to the corresponding facts stored in the KB. According to the prospective industrial scenarios where a massive number of models will be retrieved, the size of the KB is expected to be enormous and needs to be concerned. Therefore, the proposed KB is designed to be compact and represent sufficient information for reproduce the process. Only critical value of the parameters that contributes to the characteristics of individual operations and the entire disassembly process are stored. Considering IndiGolog syntax, the agent will treat each element in the Prolog facts as a fluent. According to the disassembly domain, the knowledge fact is classified

Table 5.2 Facts in knowledge base

Knowledge level	Input queries	Fact
Product-level	–Model	DSP
	–Revision version	Specific product structure
Component-level	–Model	Component location
	–Revision version	**General plan** (autonomously generated)
	–Main component instance	–Tool executing position
		–Other tool related parameters
		Add-on plan (from human assistance)
		–Customised primitive operation
		–Tool executing position
		–Other tool related parameters

into two main levels: (a) *product-level* and (b) *component-level*. The types of fact are summarised in Table 5.2. The explanation is as follows.

Product-level fact

This level represents an overview of the disassembly process in regard to the main product structure and DSP in which the agent can follow in order to achieve the goal. The agent uses the KB by searching for the corresponding knowledge according to the input queries. The proper set of knowledge can be determined based on two input queries: (a) *model* and (b) *revision*. The agent needs to recognise the model in order to follow the existing plans. The first revision version (*rev*) of the knowledge will be built if the agent has never seen this model before. On the contrary, if this model has been repeatedly disassembled for many times, the latest revision version of the knowledge will be used and modified. The proposed knowledge in the form of Prolog fact is shown in Expression (5.6).

$$dspInKb([model, rev], sequencePlan, structureType). \tag{5.6}$$

where $sequencePlan = [c1(1), \ldots, ck(i)]$ and $ck(i) = compType_k(i)$.

Afterwards, these input queries will be matched to the corresponding knowledge of DSP. The DSP is represented by a fluent *sequencePlan* which is a sequence of indexed instances of the main components to be treated. For example, $sequencePlan = [c1(1), c2(1), c3(1), c3(2), c4(1), c5(1)]$ shows an order for removing 5 types of the main component where 2 pieces of the component type *c3* are found. The $compType_k(i)$ represents the main component type-*k* with the index-*i*.

In addition, specific structure type (*structrueType*) is used to identify groups of the product models having similar characteristics, e.g. similar order of the main components. This knowledge provides extra information that helps the agent in reasoning process to achieve disassembly process more effectively. In Chap. 6, the case-study shows a clear example of the usage of this information to justify the order to treat the main component.

Component-level fact

This level stores the information regarding the treatment of a specified component. The knowledge in physical specification of the component and the related operation are given. The agent matches the corresponding fact according to three input queries, including (a) *model*, (b) *revision*, and (c) *compType*. These queries are related to the product-level fact where the current main component to be treated ($compType_k(i)$) is obtained from the *sequencePlan*.

The knowledge about the treatment process consists of three parts used for representing the location and the related operation. First, the location of the component relative to the product coordinate (*compLocation*) is used as a reference for the locating region of interest (ROI) which can be used in the process of detecting the state change. Second, the operations' details are stored as $plan_{general}$ and $plan_{addon}$ which represent the *general operation plan* and *add-on plans*, respectively. The knowledge for each main component is represented in Expression (5.7).

$$planInKb\big([model, rev], compType_k(i), compLocation, plan_{general}, plan_{addOn}\big). \tag{5.7}$$

For the general operation plan, the knowledge is initially generated by the autonomous trial-and-error process. To execute the operation plan, the agent will use this knowledge incorporating with the existing predefined general plan. The knowledge regarding the general plans is presented in Expression (5.8). Only the critical process parameters are stored in order to minimise the size of the KB. For clarification, the knowledge is represented with two types of parameters, Φ and *Γ*, which are used to reproduce the primitive actions as originally executed.

$$plan_{general} = \left[\left[\vec{\Phi}_0\right], \left[\vec{\Gamma}_0\right], \Phi_1, \Gamma_1, \ldots, \Phi_u, \Gamma_u, \ldots, \Phi_n, \Gamma_n\right] \tag{5.8}$$

Φ represents the executing position of the disassembly tool which is a critical value of the operation plan. The position relative to the main component can be indicated in a compact form according to primitive feature. Γ represents other tool's related parameters used when executing the operation. It contains a set of parameters describing the approach of the tool in detail, e.g. orientation, feed speed, etc. The primitive action can be formulated by using these parameters; for example, a rectangular cutting path at the critical depth: $\Phi = rect(x_1, y_1, x_2, y_2, z_c)$ with tool orientation *m* with feed speed *s*. With the cutting contour rectangle operation *cutContour*(), the primitive action is $cutContour(x_1, y_1, x_2, y_2, z_c, m, s)$.

For the added-on plan, the knowledge is generated according to assistance given to resolve specific issues happened during the process. The primitive actions with the critical parameters are stored directly as *primAction*(Φ, Γ) as shown in Expression (5.9).

$$plan_{addOn} = \left[primAction_1(\Phi_1, \Gamma_1), \ldots, primAction_n(\Phi_n, \Gamma_n)\right] \tag{5.9}$$

5.5.2 Learning

The learning process collects the knowledge contributing to the successful disassembly process of a particular model of the product. This knowledge is used for reproducing the same disassembly outcome for the previously seen models. The same disassembly outcome will be achieved based on assumption that the main components are treated in the same order; meanwhile the operations for each main component are assumed to be *order-independent*. Therefore, some variations occur at the operation level and there is a possibility of improvement. Essentially, in the learning process, all operations that have been performed need to be recorded even where the state has not been changed immediately after the operation has finished. The assumption is that all performed operations have contributed to the upcoming state change unless there is *a proof of redundancy* which can be considered by the revision process (see Sect. 5.5.4). The normal form of learning is performed only the first time that the new model has been recognised. Afterwards, the learning will occur in the form of revision. In this section, two normal forms of learning, i.e. (a) *by reasoning* and (b) *by demonstration*, are described.

Learning by reasoning occurs throughout the trial-and-error process. The knowledge is obtained as the agent conducts the process autonomously according to the basic behaviours of the general plans and operations. According to the KB, all facts will be generated except the fact regarding $plan_{addOn}$ which is generated by the learning by demonstration. Each type of the fact is generated in a different stage of the plans and operations.

For the *product-level knowledge*, all elements will be obtained after the entire disassembly process has been achieved. First, the component is continuously added to a list of the fluent *sequencePlan* once the new main component has been detected. Second, the *structure type* can be indicated after classification of the structure types if needed. The classification process can be different according to the variations found in each product. Currently, this classification strategy needs to be predefined by the user before starting the process. However, in the future, to increase flexibility of the system, an ability to add new structure types during the process is possible by modifying the CRA using *learning by demonstration* strategy [14].

For the *component-level knowledge*, the critical value of parameters entailing the state change or the last action before proceeding to the next operation plan need to be learned due to the aforementioned discussion regarding the redundant operations. This information will be learned after the execution monitoring. If the state has changed, the current plan and parameters are recorded as the critical values and proceed to the next state. If the state has not changed, the current parameters are recorded and the agent will try alternative parameters or operation plans. In this case, the information will be revised and only the final values of the parameters of each operation plans will be kept.[3] After all available general plans have

[3] The revision strategy is similar to one that is done for the human demonstrated primitive actions sequence. It is clearly explained in the section of learning by demonstration where the inaccurate detected location has been addressed [see Eqs. (5.14–5.15)].

been executed or the state has been changed, all required parameters for $plan_{general}$ will be stored in the KB file.

5.5.3 Learning by Demonstration

Tance is involved in this type of learning. The user demonstrates actions which are exogenous actions to address the unresolvable condition. The actions can be a sequence of primitive actions when the agent has struggled to remove the component after all available autonomous operations have been attempted. In addition, the action can be used to change original belief of the agent in case of initially misperception. Theoretically, the assistance is required only for the first time in disassembling the unknown model in order to handle complex situations. The agent supposes to learn from this demonstration and autonomously carries out the entire process the next time this model is found. However, according to the physical uncertainties in the product and the process, minor assistance may be needed in a few subsequent disassembly processes to resolve remaining minor uncertainties.

The problems are associated with the component level and the operational level. The unresolved conditions typically result from the imperfection of the sensors in various detection processes. The problem can be classified into three types, including (a) *false detection of components*, (b) *inaccuracy in localisation*, and (c) *non-detectable components*. Proper types of demonstration need to be given to resolve these problems after the autonomous operations have been finished. These conditions are summarised in Table 5.3.

Table 5.3 Unresolvable conditions to be addressed by human assistance

Level	Unresolvable conditions	Type of problem			Sensor	
		False detection by vision system	Localisation inaccuracy	Non-detectable	Vision	Force/tactile
Component	Existence of components	•			•	
	Location of components		•		•	
	Detection of state change	•	•		•	
Operation	Tool executing position		•	•	•	•
	Remaining established connections	•	•	•	•	
	Physical collision with outer objects			•	•	•

False detection

The false detection directly relates to the vision system. The false positive or negative detection affects the reasoning process about the existence of the component and state change. After the information has been sensed, *belief* about the condition of the current disassembly state will be initially created in the agent. This wrong belief can result in logical and physical fault in the subsequent operations.

The *false positive outcome* results in executing the actions that do not suppose to be executed. For the main and connective components, excessive damage on the components possibly occurs due to improper treatment of the component. Subsequently, since the agent believes that a certain component exists, it will try many ways to remove the component which does not exist in realty. This trial can harm the product and waste time. For the state change, the agent will realise that the removal of the component has been successful while it does not happen in reality. Therefore, logical understanding of the main structure of the product will be wrong. The user can give action in Eq. (5.10) to manually indicate the state change resulting in skipping of this component.

$$stateChanged = true \tag{5.10}$$

On the contrary, the *false negative outcome* results in an ignorance of necessary actions needed to be executed. For the main components, they will not be treated properly and will not be removed. The disassembly state related to this main component will be skipped (human operator gives action in Eq. (5.11)). Therefore, this component will still attach to the product and preventing the system to access the components in the subsequent state. The perception about this main component in main product structure will be incorrect also. For the connective components, it is high possibility of having false negative result due to their small size and occlusion. The fault detection will lead to the connections untreated; consequently, the main component will be unsuccessfully removed. For the change of disassembly state, the agent will not realise that the component has been already removed. Consequently, it will keep trying to remove the non-existence component which leads to infinite loop of trials since substantial change will not be able to take place.

$$action = skipComponent \tag{5.11}$$

Inaccuracy in localisation

Localisation issue takes place after the components have been properly detected (true positive). Regarding the location of components, the location is determined by detection algorithm of the vision system. Inaccuracy is associated with the performance of the detector used and the EOL condition that can vary according to the usage phase. Therefore, location error within an acceptable range is likely to occur due to these uncertainties. The location of the component is an original belief that is used as a reference ROI for determining state change. In case that significant error occurs, success of the plan will be inaccurately measured. Therefore, assistance needs to define the new accurate component location (see Eq. (5.12)).

$$action = compLocation(< component\ location >) \tag{5.12}$$

The tool executing position involves at the operation level. It is a consequence of the reasoning process of the location of the component and the corresponding operation plans. Inaccurate executing position will result in a failure to remove the component. The assistance is needed in order to give extra operations to compensate this position error in addition to the agent's actions. Since the operation plan generally composes with a sequence of primitive actions, the assistance must be given in the same form as in Eq. (5.13).

$$action = \left[primAction_1;\ primAction_2;\ \ldots\ primAction_n\right] \tag{5.13}$$

Each executed action needs to be stored in the KB individually due to the specificity. Therefore, the size will be extended according to the number of times assistance given. To reduce this size, a strategy to refine the critical value from a sequence of actions producing similar result should be implemented. Then, only a critical action will be kept in the KB. This is typically applied to the destructive operations where the change, i.e. damage, is explicit on the objects. For example, a demonstration that repeatedly drill at position (x_0, y_0) incrementally deep down to the critical depth. Multiple vertical drills at various depth z_0, z_1, ... , z_k have been demonstrated along the way. Given that $z_0 < z_1 < \ldots < z_k$, only the primitive action with the deepest cut z_k should be kept. The original and the refined action sequences are shown in Eqs. (5.14–5.15), respectively.

$$action_{original} = [cutPoint(x_0,y_0,z_0), cutPoint(x_0,y_0,z_1), \ldots cutPoint(x_0,x_0,x_k)]. \tag{5.14}$$

$$\textbf{if}\ (z_0 < z_1 < \cdots < z_k)\ \textbf{then}\ action_{refined} = cutPoint(x_0,y_0,z_k) \tag{5.15}$$

Non-detectable components

This category deals with non-detectable objects which are not expected to be detected or incapable to be detected by the sensors. In the unknown environment according to the variations in products, disassembly operations depend on the sensing capability rather than the prior knowledge. If the environment cannot be fully described by the sensors, the agent will not be able to execute proper actions to achieve the goal of disassembly process. The non-detectable connections and the physical collision with other object are the major problem. Difficulty in determining the proper tool executing positions is a consequence of these problems. Therefore, a sequence of primitive action as in Eq. (5.13) must be given as assistance.

For the connections, one of the challenging problems is to the detectors that is capable of recognise and localise particular types of the connections. According to the explanation in Chap. 4, the connections are categorised into three types, including (a) *quasi-component*, (b) *virtual-component*, and (c) *non-components*. The quasi-components are highly possible to be detected since they are visually observable, e.g. screws, plastic rivet, snap-fits, cable, etc. However, the detection rate can be low in certain circumstances if they are partially occluded or totally hidden by other objects. On the contrary, the virtual components are more difficult to be detected due to the size and observability, e.g. weld and soldering. For the non-component connections, these are not able to be detected visually since there is no extra connective components involve, e.g. press-fit, glue, mating, etc. In summary, the connections are not able to be properly detected and disestablished in most case. Consequently, the corresponding main component will be irremovable. Therefore, the assistance should be provided in order to locate and remove them.

In regard to the physical collision, when the robot tries to approach an object to be executed, the accessibility can be limited by physical constraints. It is possible that the disassembly tool or part of the robot will crash with other part of the product during the operation. This circumstance can be prevented if the product and the movement path of the automation are accurately modelled and planned. Therefore, force-tactile sensors are typically used in addition to the vision system to enhance the perception of the system. However, this modelling and planning can become more complex and challenging due to the computing resource with respect to the real-time internal and external information. Therefore, if the robot cannot resolve the collision while accessing the object, the human operator should demonstrate tool paths for a proper movement.

5.5.4 Revision

The revision process aims to optimise the disassembly process of the known model by removing redundant operations and reducing the size of the operation plan set in operation-level fact that have been learned. Therefore, the efficiency of the process in term of time consumption will increase. The reduction will be done by retracting the redundant operations that do not contribute to the removal of the main component. This process is expected to be carried out autonomously.

To find out the redundant operations, the treatment process of each main component needs to be repeated by executing the operations in a different order. The process with this variation is expected to be repeated multiple times to obtain the optimal set of operation plans. The smallest subset of the operations that achieve the disassembly goal can be found out by execution monitoring. The subset is obtained by retracting all redundant steps based on assumption that all performed operations contribute to the upcoming state change unless there is a *proof of redundancy*. This redundancy can be proved by not executing some of the previously recorded operations. If the main component is still successfully removed by executing the existing set of operations, the removed operation can be concerned redundant. Retraction of the redundant operation plans will be done subsequently. The KB will be repeatedly revised based on the latest version once the identical known model has been seen. Finally, the revised version of the facts will be stored in the KB.

An example for redundancy is shown in Fig. 5.13 where the final material removed from cuts are related to only *op*(i,2) and *op*(i,3) but all of them are recorded in the learning process. The *op*(i,1) can be considered a redundancy since its cutting area can be covered by another two operations. Therefore, the operation *op*(i,1) has been proved as redundancy which can be retracted (see Fig. 5.13c) or unnecessary to be recorded at the first place.

The retraction strategy focused on the knowledge associated with the general plans since they are created autonomously by trial-and-error. Therefore, it is high potential that the redundancy is generated. Unlike the added-on plans, operations demonstrated by the human operators are expected to be more specific and necessary for the components removal. It can be assumed that none of the redundant operations occur in this assistance. Therefore, only the general plans will be taken into account.

General approach

A diagram in Fig. 5.14 illustrates a general strategy for finding out the redundant operations by repeating the selected operations from previous process. Initially, to achieve removing a component in the first revision, the operations were executed as this order: *op*(*c*,1) → *op*(*c*,2) → *op*(*c*,3) → *op*(*c*,*H*). In the second revision, the executing order will change to many different cases. The *op*(*c*,*H*) which is a

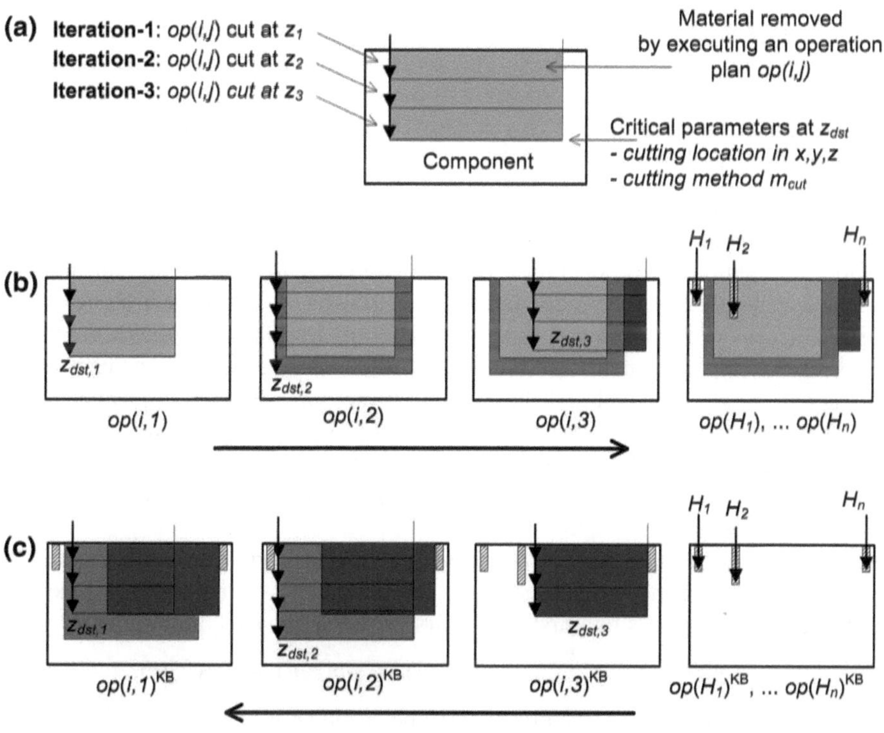

Fig. 5.13 Cutting operations in learning and implementation

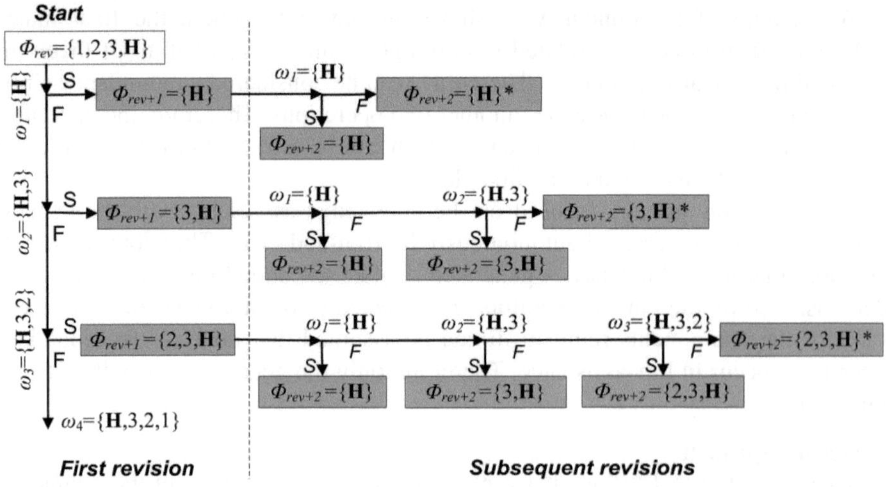

Fig. 5.14 Strategy for retracting the redundant operation

sequence of added-on plans will be executed in any cases and only one operation plan will be skipped in each process. If the component is successfully removed, it can be concluded that this operation is redundant and can be retracted from the KB. In subsequent process, one more operation will be taken out from the remaining valid operation. In case that the removal is failure, another operation plan will be considered. This process will be repeated until all valid combinations of the plans have been tested. This strategy is able to reduce the testing times by reducing the search-space.

Simplified approach: an example for heuristic approach

The aforementioned general approach expects to comprehensively check all valid combination of the operation plans. However, enormous time consumption due to the number of repetitions needed is a drawback. In this section, a simplified retraction strategy using heuristic approach is described. In this example, the order of the operation plans is designed to be directly related to the success rate of the component removal and the impact on the component, e.g. physical damage. Initially, the lower indexed operation has low impact but low success rate; whereas, the higher indexed operation has high success rate but high impact. The assistance added-on plan is considered as the most specific actions and will not be retracted as always.

In this case, the success of removal is the highest priority of the heuristic strategy. Therefore, to achieve the time constraint, the agent will try executing the plan with the highest success rate first. In the first revision, the executing order is

$$op(c,1) \rightarrow op(c,2) \rightarrow op(c,3) \rightarrow op(c,4) \rightarrow op(c,H).$$

In the subsequent revision, the process will be carried out in a reversed direction:

$$op(c,H) \rightarrow op(c,4) \rightarrow op(c,3) \rightarrow op(c,2) \rightarrow op(c,1).$$

With the assessment from execution monitoring, the retraction can be simplified by retracting all plans with lower indices than the current operation without considering the intermediate plans. For example if the component has been removed after executing *op*(*c*, *H*) and *op*(*c*, 4), the remaining plans *op*(*c*, 1)–*op*(*c*, 3) will be retracted without considering an effect of reordering among these plans as done in the general approach. In Procedure (5.16), the retraction for component c_i of the sample "model" is shown.

```
Proc generalPlanRetraction
  < verDsp = j >                          // current revision j + 1 use previous verDsp = j
  executeAddOnPlan(model, c_i);           // execution the custom plan sequence in KB
  k = n;                                  // start with the highest plan index-n : k = n
  while ¬stateChanged ∧ k ≥ 1 do
    op(c_i, k);                           // execute plan-k in KB
    checkStateChange;
    if stateChanged then                  // if state changed, retract the lower index plans
      retractGeneralPlan([op(c_i,1), ...., op(c_i,k−1)]);                    (5.16)
    else k = k − 1;                       // proceed to the lower index plan
    endIf
  endWhile;
  if ¬stateChanged then callUser endIf    // if fail, add more custom plans and learn
  writeToKb(planInKb([model, j+1], c_i,
        compLocationKb', plan'_general, plan'_addOn))   // record the revisied verDsp to KB
endProc
```

In summary, this simplified approach can find and retract the redundant operation more quickly than the general approach. A drawback is that the redundant operation plan within the intermediate plans cannot be retracted in any case. However, this simplified approach is effective enough to improve the process performance given that the operation plans are in a proper order. This concept has been proved by the experiment of the case-study which is explained in the next chapter.

Cognitive robotic agent's self modification

Previously, the revision is done on the KB with is a separated part from the cognitive robotic agent. In this case, the cognitive robotic agent will modify its own program in order to revise the program structure which represents its behaviour. The agent will be more flexible to learn and generalise the operation plans from the disassembly process. Eventually, the agent is expected to be less associated with the model-specific knowledge. With the generalised plans, the agent is expected to be capable of disassemble previously unseen models of product more effectively without assistance. According to the learning by demonstration proposed by Braun [14], the version-space is used for considering the self-modification of the Golog program according to the demonstrations and the existing program. As a result of the learning process, relations between input and output will be formulated. The input is sensible information of a type of component. The outputs are the specific non-detectable connections needed to be disestablished and the corresponding operations. Afterward, the agent will be able to perform proper operations as per information perceived.

Two major challenges are involved with this approach. First, to accurately identify relations is one of the major challenges of this concept. However, these relations have not been discovered yet; for example, there is no relation between the required cutting destination of a hidden snap-fit underneath the middle area of a back cover and the location of the back cover's border. Second, the cognitive robotic agents will be able to evolute in totally different ways according to the training samples. In actual disassembly scenario, the training set is variable and unpredictable due to the EOL product returned. Therefore, after the agents have been trained with a certain number of models, to merge the agent together in behavioural level can be complicated. Currently, his concept has not been successfully implemented by the time this book is being written.

5.6 Implementation Procedure

In order to integrate this cognitive robotic module in disassembly automation, the automation needs to satisfy the requirements regarding the functions needed by the agent. The requirements of each module is summarised in Fig. 5.6. Regarding the cognitive robotics module, the agent is designed based on the cognitive behaviour presented in this chapter. In addition, the functionality and performance of the supporting modules are taken into account. The cognitive robotic module will be created considering the integration procedure of the entire system (see Sect. 3.5). The procedure is summarised as follow:

- Analyse the product,
- Analyse the disassembly requirements;
- Design and assess performance of the functions needed VSM and DOM;
- Create disassembly domain;
- Determine the cognitive robotic functions;
- List the actions, fluents, precondition, effects required for each modules;
- Program the behaviours using Procedure due to complex actions; and
- Connect the modules together via actions and fluents.

In prior to create the cognitive robotic module, the product needs to be analysed and the desired disassembly strategy needs to be decided. The supporting modules also need to be assessed. For the cognitive module, the disassembly domain should be created first; therefore, the overview of the process can be seen more clearly. After that, the requirements in the disassembly process and operation will be implemented by the cognitive behaviours. The four cognitive functions are needed to be designed for performing these tasks. From the programming perspective, the operations are needed to be represented in the form of actions, fluents, preconditions, and effects. These relations can be written as successor state axioms and the complex actions can be defined as Procedures. These elements related not only to the CRM but also VSM and DOM. As a result, the communication among these modules is established and ready for performing the disassembly process.

5.7 Conclusions

The concept of cognitive robotics has been applied to many fields of research. This chapter presents the implementation of cognitive robotics in disassembly context. The main objective is to increase the flexibility and robustness of the system. Basically, the cognitive ability allows the system to handle incomplete knowledge of the world in more effective way. As a result, the problem regarding uncertainties in the product and process can be resolved.

Firstly, the information in disassembly process is needed to be represented in the form that the cognitive robotic agent can understand and utilise. KRR is involved in structuring the disassembly domain which represents the state of disassembly. In this case, the disassembly domain consists of three levels: the product structure, disassembly operation plan, process parameters. The CRA controls the system according to the basic and the advanced behaviour. For the basic behaviour, the agent can reason about the disassembly process and schedule the actions for each operating modules in the system. To make the process more robust, execution monitoring is used for assessing the outcome of the operations. When the operation fail, the alternative operations and parameters will be used in order to complete the goal. In addition, thank to the advanced behaviour, learning and revision strategy make the process becomes more efficient after repeating disassembly the previously seen model.

In conclusion, it is clear that the majority of the uncertainties in the disassembly process can be addressed by the cognitive robotic functions. However, in the actual process, the performance of the system also relies on other operating modules in which imprecision is present. These uncertainties occurred from those sources may introduce a new problem that is unable to be resolved by the cognitive robotic. Therefore, the assistance is still needed for handling unresolved problems. However, due to the learning capability, the human intervention is expected to be reduced after some revision. The system will be fully automated eventually. In the next chapter, the validation of this concept will be clearly described through the case-study.

References

1. Moreno RA (2007) Cognitive robotics. http://www.conscious-robots.com/en/conscious-machines/the-field-of-machine-consciousness/cognitive-rob.html. Accessed Oct 2009
2. Müller VC (2012) Autonomous cognitive systems in real-world environments: less control, more flexibility and better interaction. Cogn Comput 4(3):212–215
3. Levesque H, Lakemeyer G (2007) Cognitive robotics. Handbook of knowledge representation: foundations of artificial intelligence. Elsevier, Amsterdam, pp 869–882
4. Zaeh M, Lau C, Wiesbeck M, Ostgathe M, Vogl W (2007) Towards the cognitive factory. Paper presented at the international conference on changeable, agile, reconfigurable and virtual production (CARV), Toronto, Canada
5. Beetz M, Buss M, Wollherr D (2007) Cognitive technical systems—what is the role of artificial intelligence? Lecture notes in computer science (including subseries lecture notes in artificial intelligence and lecture notes in bioinformatics), vol 4667, pp 19–42

6. Bannat A, Bautze T, Beetz M, Blume J, Diepold K, Ertelt C, Geiger F, Gmeiner T, Gyger T, Knoll A, Lau C, Lenz C, Ostgathe M, Reinhart G, Roesel W, Ruehr T, Schuboe A, Shea K, Stork Genannt Wersborg I, Stork S, Tekouo W, Wallhoff F, Wiesbeck M, Zaeh MF (2011) Artificial cognition in production systems. IEEE Trans Autom Sci Eng 8(1):148–174 (art no 5524092)
7. Vongbunyong S, Kara S, Pagnucco M (2012) A framework for using cognitive robotics in disassembly of products. In: Leveraging technology for a sustainable world—proceedings of the 19th CIRP conference on life cycle engineering, pp 173–178
8. McCarthy J (1963) Situations, actions, and causal laws. Technical Report Memo 2, Stanford Artificial Intelligence Project. Stanford University
9. De Giacomo G, Lespérance Y, Levesque HJ, Reiter R (2001) IndiGolog-OAA interface documentation. http://www.cs.toronto.edu/~alexei/ig-oaa/index.htm. Accessed 16 July 2011
10. Lapouchnian A, Lespérance Y (2002) Interfacing IndiGolog and OAA—a toolkit for advanced multiagent applications. Appl Artif Intell 16(9–10):813–829
11. SWI-Prolog (2010) SWI-Prolog. http://www.swi-prolog.org
12. Ghallab M, Nau D, Traverso P (2004) Automated planning theory and practice. Morgan Kaufmann Publishers, Elsevier, San Francisco
13. Vongbunyong S, Kara S, Pagnucco M (2014) Learning and revision in cognitive robotics disassembly automation. Robot Cim-Int Manuf. http://dx.doi.org/10.1016/j.rcim.2014.11.003
14. Braun A (2011) Programming by demonstration using the high-level programming language Golog. Diploma, RWTH Aachen University and The University of New South Wales

Chapter 6
Implementation and Case-Study

Abstract The automated disassembly system consists of three main operation modules presented in Chaps. 3–5. In this chapter, procedures for system integration and the development of each module are described in detail, with regard to a cognitive robotic disassembly system designed for disassembling a case-study product, the Liquid Crystal Display (LCD) screen. Finally, the performance of the proposed system is validated by experiments which demonstrate the flexibility and robustness of the system.

6.1 Implementation Overview

Economic feasibility is a major concern in the actual implementation of a disassembly system. This is the main reason why automation is proposed to replace the high-cost human labour. The main objective is therefore to develop a low-cost automated system that is both flexible and robust. The system needs to be capable of handling uncertainties and variations in product and process. To develop the automated disassembly system, the procedure is summarised into the following 6 main steps:

- Product analysis and disassembly strategy,
- System framework,
- Cognitive robotic module,
- Vision system module,
- Disassembly operation unit module, and
- Validation and performance testing.

Firstly, the product to be disassembled is examined to obtain the main product structure and components (see Sect. 6.2). This information is used to create the disassembly domain used by the cognitive robotic agent. The strategy to disassemble this product is needed to be identified (see Sect. 6.3). The information about the components is used for designing the functions of the vision module system and disassembly operating module.

S. Vongbunyong and W.H. Chen, *Disassembly Automation*, Sustainable Production, Life Cycle Engineering and Management, DOI 10.1007/978-3-319-15183-0_6

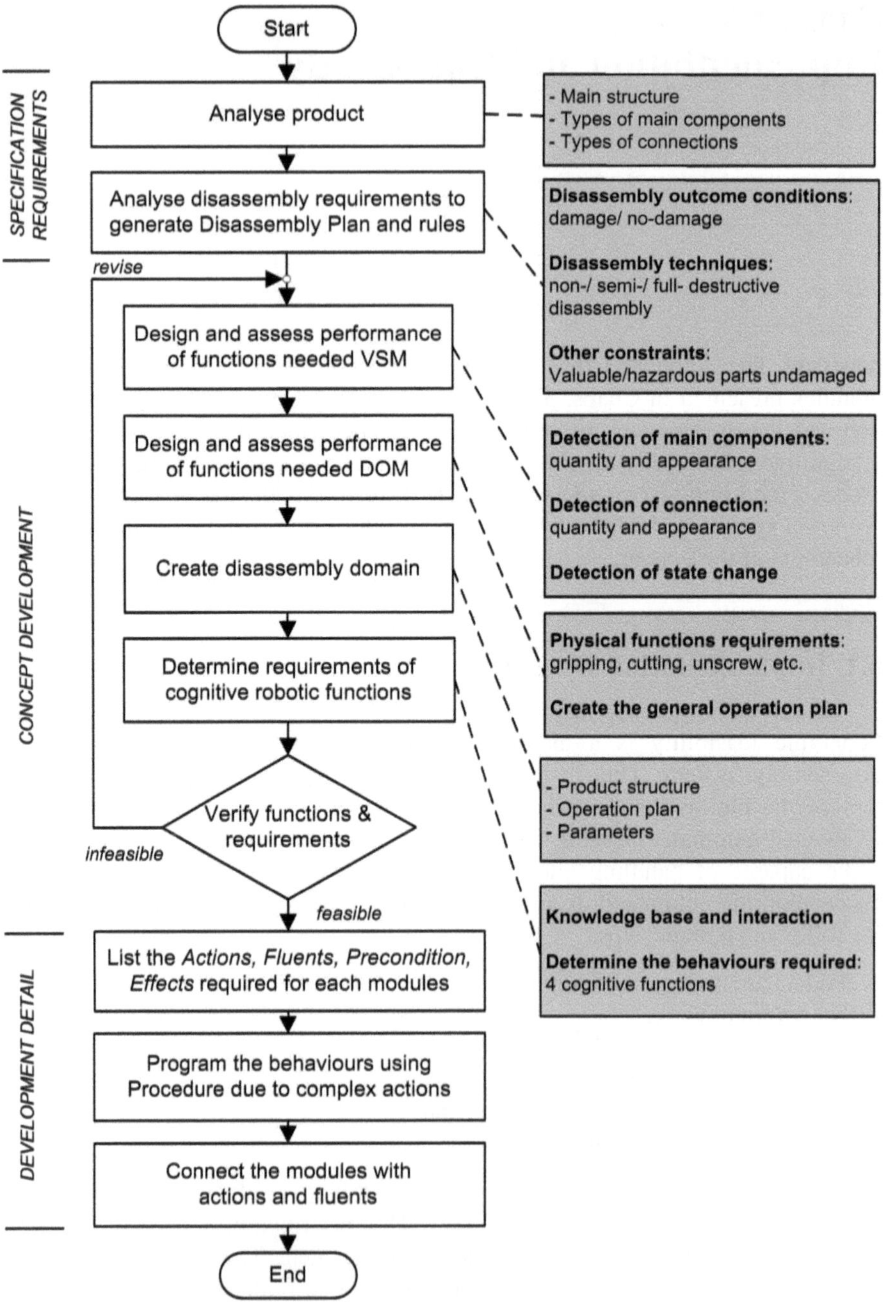

Fig. 6.1 Flowchart for developing cognitive robotic disassembly automation

The system framework (Sect. 6.4) provides an overview over the various control levels and interactions among the operating modules. Each operating module is designed to address the uncertainties in the disassembly process (see Sects. 6.5–6.7). Regarding the concept of cognitive robotics, the activities and behaviour of each module are subsequently represented in the form of actions and fluents. Finally, the performance of the system is validated according to the objective of the system. These procedures are summarised in Fig. 6.1 and explained in detail in the following sections.

6.2 Product Analysis

6.2.1 Case-Study Product: LCD Screens

LCD screens have dramatically replaced Cathode Ray Tube (CRT) monitors over the past decade. More than 120 million units of LCD screens were sold worldwide in 2008 and used in approximately 90 % of desktop computers in 2010 [1]. According to an earlier prediction, the sales of LCD screens would be US$80 billion in 2012—approximately four times higher than that of other types of monitors [2]. Therefore, the number of disposals is continuously increasing. In Germany alone, it is expected that more than 4,000 tons of LCD screen monitor will be disposed of by 2012 [1]. The environmental impact of this product is significant and increasing. EOL treatment needs to be considered.

6.2.2 End-of-Life Treatment of LCD Screen Monitors

Treatment of EOL LCD screens is considered according to the European Directive on WEEE [3]. This directive indicates the recovery rate as well as hazardous substances that need to be taken into consideration: both aspects need to be considered in the design of the disassembly process. The minimum threshold requirements by weight are 75 % for material recovery and 65 % for recycling and reuse.

This study started with the investigation of a number of LCD screens, in order to gauge their material composition and the commonly-found components. The majority of the overall weight is contributed by the main components, i.e. lightbox casing, PCB mounting panel, back cover, and PCBs [2]. According to the material, the distribution is shown in Fig. 6.2 [4]. The target of 78 % material recovery by weight is achievable if material fractions, i.e. ferrous metal and Halogen-free plastic, are separated before recycling.

Regarding the treatment of hazardous substances indicated by the directive, the removal of three potentially hazardous components must be taken into account: (a) the Cold Cathode Fluorescent Lamps (CCFL), (b) the LCD glass panel, and (c) PCBs [2]. The disassembly of CCFLs (a) is difficult due to its fragility and strong

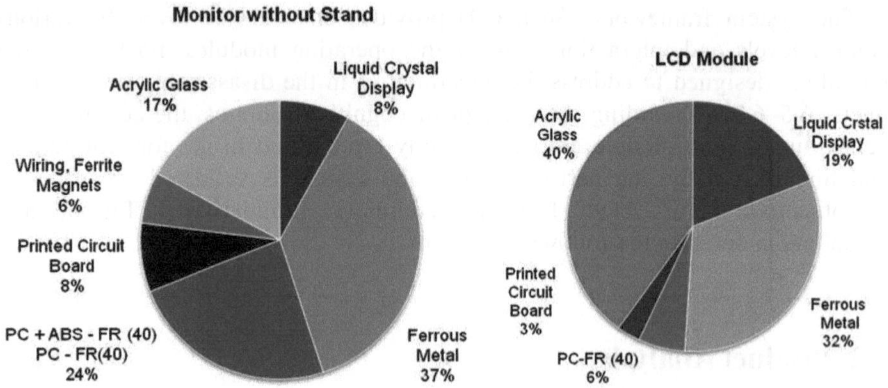

Fig. 6.2 Distribution of the material in LCD screen (adapted from [4])

connection with other components in the LCD module. Special treatment is required due to the small amount of Mercury contained in CCFLs. The amount of Mercury varies in different screen with approximately 4.8 μg per screen[1] on average [4]. The LCD glass panel (b) is a potential environmental risk. Any LCD larger than 100 cm^2 must be removed from any WEEE. Therefore, the LCD module must be disassembled to retrieve LCDs. It is usually carried out destructively [1]. Third, the PCBs need to be separated since they contain several types of metal and thermoplastic material which is difficult to recycle. Any PCB larger than 10 cm^2 must be removed from the WEEE [2].

It is clear that component separation is important in the EOL treatment process for LCD screens. The disassembly process needs to be designed well to be efficient and satisfy the constraints regarding the hazardous components. However, achieving economic feasibility via manual disassembly is still a challenging problem. It was decided to implement selective disassembly to the module level. The LCD module can then be further destructively disassembled but CCFL and LCD should be secured.

6.2.3 Structure Analysis and Components

The basic structure of the product is needed in formulating the disassembly domain used by the cognitive robotic agent. Information regarding the components is needed for developing the functions for visual detection and operation plans.

[1] Estimate that 250–480 kg of Mercury potentially found in 80 million LCD screens are expected to be disposed by year 2010.

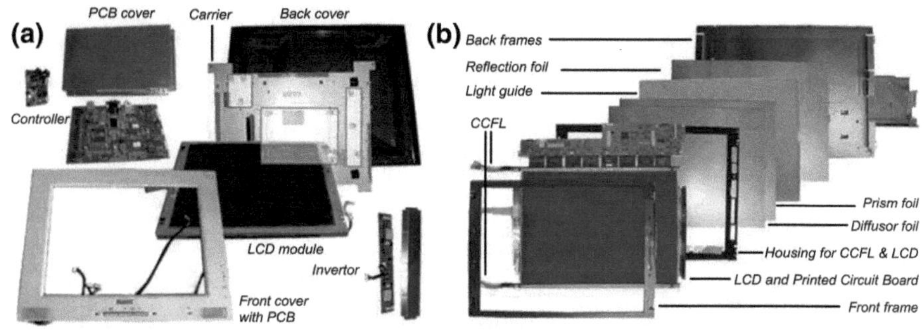

Fig. 6.3 Structure of LCD screen [5]. **a** Modules in LCD screens. **b** Components in LCD a module

A comprehensive study of the structure and components of LCD screen are reported in many research works [2, 5]. A common structure can be presented in two levels: (a) *module level* and (b) *component level* (see Fig. 6.3). Disassembly at the module level can be examined by the selective disassembly approach. This product consists typically of 6 types of main components: a back cover, PCB covers, PCBs, a carrier, an LCD module, and a front cover. Three types of PCBs are commonly found, i.e. a power supply—inverter board, a switch board, and a controller. This order of the component is from the front to the back of the screen. This general structure of LCD screens is similar among the different manufacturers. However, the location of the components, e.g. PCBs, screws, cables, can be significantly changed even within manufacturer families. With respect to the component level, the LCD module can be further disassembled into 9 components (see Fig. 6.3b).

As the disassembly system is aimed at material recovery, this solution considers the concept of component and structure in a slightly unusual way. Rather than the specific identification of components, only the component type needs to be determined. Such a definition is beneficial in this context because the system is then able to immediately begin operation on previously-unseen models. A more relaxed definition of product structure is also adopted, whereby only the attachment direction and relationships are taken into account. This solution exploits the advantages of the cognitive functions and operating modules, in particular the vision system. Figure 6.4 describes the product at the module level under this scheme.

The product structure can be classified into two types according to assembly direction: *Type-I* and *Type-II*. This classification is based on the configuration of components PCBs, PCB cover, and carrier. The PCB cover is a separate part in the Type-I structure, whereas it is integrated with the carrier for Type-II. In addition, Type-I can be further categorised into two sub-classes according to the material and appearance of the PCB cover. Type-Ia is visually distinguishable from the carrier by the vision system while Type-Ib is not. The CRA incorporated

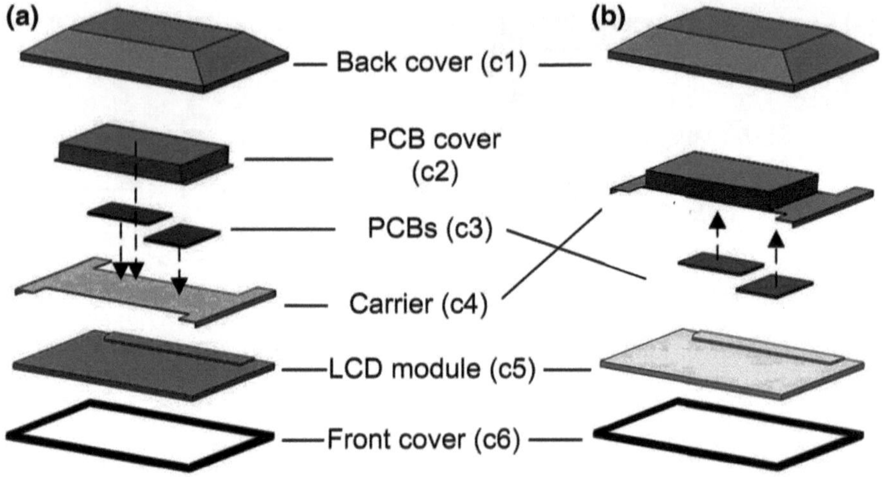

Fig. 6.4 Product structure of LCD screens and the main components. **a** Type-I. **b** Type-II

with physical disassembly operation is used to differentiate these two sub-classes. LCD screens typically consist of six types of main components and four types of connective components denoted by '*c*' and '*cn*', respectively. A list of components is as follows:

- Back cover (*c1*);
- PCB cover (*c2*);
- PCBs (*c3*);
- Carrier (*c4*);
- LCD module (*c5*);
- Front cover (*c6*);
- Screws (*cn1*);
- Snap-fits (*cn2*);
- Electric cables (*cn3*); and,
- Plastic rivets (*cn4*).

The variation of the structure and appearance of the components are the major challenges in the automated disassembly. Examples of the liaison diagrams of LCD screens with each structure types are shown in Fig. 6.5 and an example of a very complex structure is shown in Fig. 6.6. Using traditional approaches for effective automated disassembly, all relevant information regarding the product and process would typically be specified a priori. This information is model-specific and unfortunately often unavailable due to the confidentiality of the manufacturers' design. Instead, the recognition of component by type and the principle of cognitive robotics are applied to resolve the variations and uncertainties.

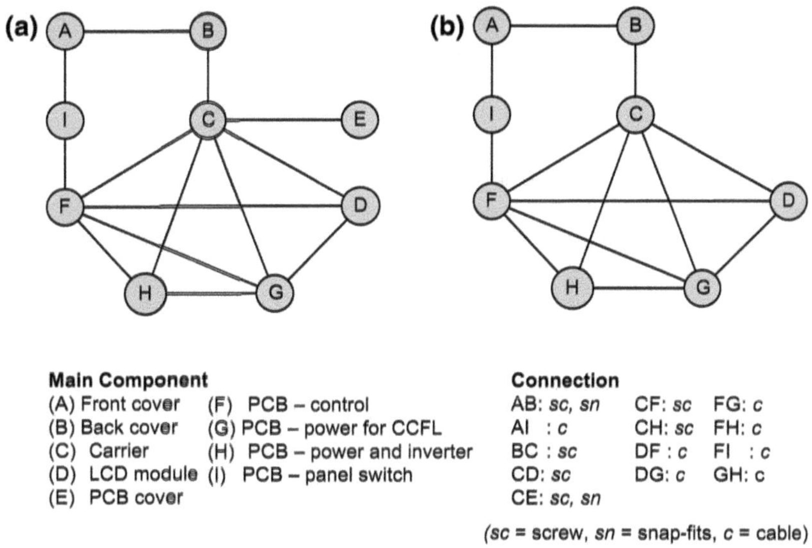

Fig. 6.5 Liaison diagram. **a** Type-I. **b** Type-II

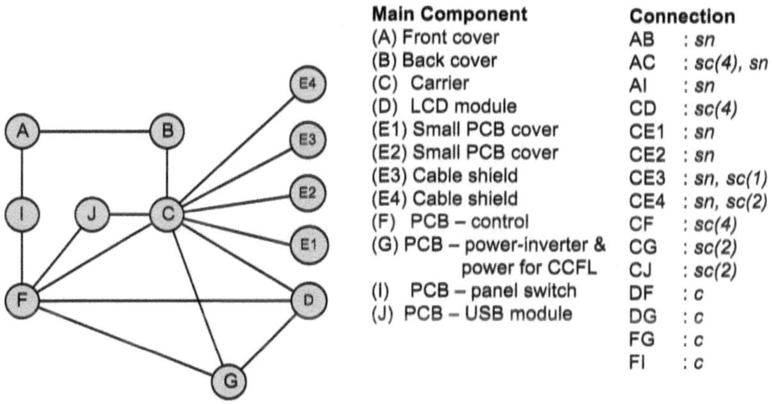

Fig. 6.6 Example of a complex structure LCD screen

6.3 Disassembly Requirements

Disassembly is a low-profit activity but nonetheless important for EOL treatment. Therefore, economical feasibility is a major concern. The automated system needs to be low-cost but flexible and robust enough to deal with a variety of the samples. According to the operation, (semi-)destructive disassembly is suggested due to its robustness and flexibility [6]. However, the damage to the disassembled parts is a major drawback. Therefore, this approach is appropriate for recycling but not suitable for reuse or remanufacture. Regarding disassembly strategy, it is difficult to achieve economically feasible full disassembly due to the decreasing value returned from time consumed. As a result, the LCD screens are selectively disassembled to the module level [5]. In addition, valuable components, as well as the hazardous constituents according to the WEEE directive, need to be considered. Breakage of CCFLs results in leakage of mercury. Hence, the LCD module will not be further disassembled to avoid damage to the CCFLs. The PCBs are crucial for removal. In summary, the automated system is designed based on the following requirements.

- Low-cost design with sufficient flexibility and robustness;
- (Semi-) destructive disassembly;
- Selective disassembly; and,
- Special requirements for the hazardous and valuable parts.

6.4 System Overview

The control architecture of the system is designed based on the cognitive robotic architecture presented in Sect. 5.3.2 with respect to the simplified scope of the case-study product. The detailed configuration of the system in regard to level of control and operating module is illustrated as a schematic diagram in Fig. 6.7.

6.4.1 Levels of Control and Operating Modules

The system can be considered in two complementary perspectives: the level of control, and the operating module. The level of control specifies the degree of automation and level of data abstraction of a given process; the operating module is the functional system in which the process belongs. One operating module can consists of multiple levels of control depending on the behaviour of processes needed for carrying out the module's functions.

Levels of Control

The levels of control relating to the individual operation modules are explained in detail in earlier chapters. This chapter only provides an overview such that the connections and information flows between the modules can be clearly seen. The system consists of three levels of control: (a) *high-level*, (b) *mid-level*, and (c) *low-level* (see Fig. 6.7).

The *high-level* controls the top-level behaviour of the system and planning of the disassembly process. Commands and information transferred to other levels are in abstract form. The cognitive robotic module operates at this level.

The *mid-level* is used to manage the information exchange between the high-level and the low-level. It is at this level where data is interpreted into information

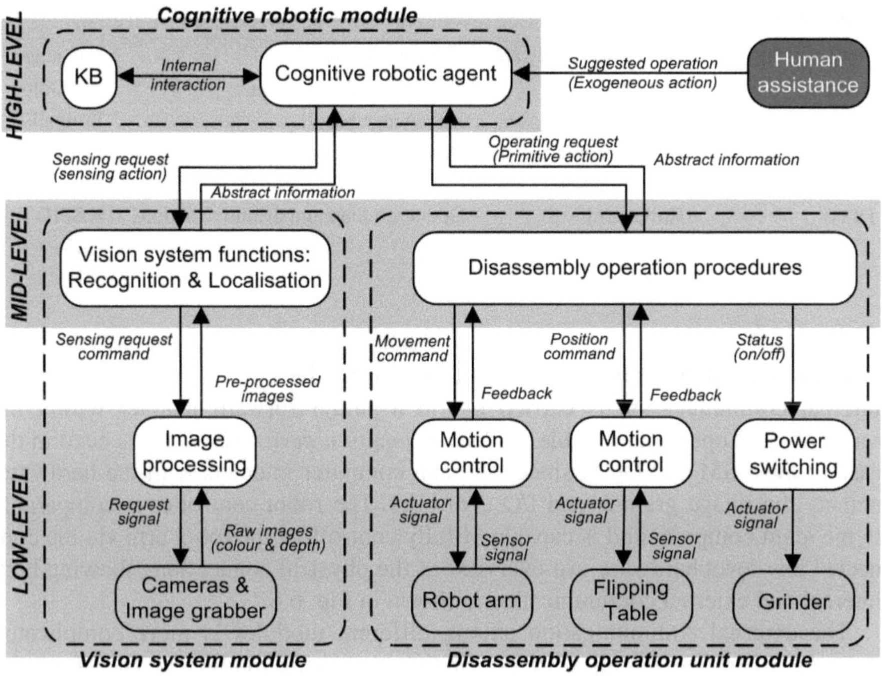

Fig. 6.7 System architecture—levels of control and operating modules

and abstract information into realisable actions. The detection functions of the vision system and the operation procedures of the disassembly operation unit modules operate at this level.

The *low-level* deals with machine-level hardware operation and signal processing, including sensor-actuator motion control and image acquisition and pre-processing.

Operating Modules

The system consists of three modules, which are designed based around specific functionalities and operate independently from each other. The operating modules and the associated hardware and functions is summarised in Table 5.2.

The *cognitive robotic module* (CRM) consists of cognitive robotic agent (CRA) and knowledge base (KB). The high level planning and interaction with the human operator are its main functions.

The *vision system module* (VSM) consists of the detection functions and two cameras—a colour camera and a depth camera.

The *disassembly operation unit module* (DOM) is responsible for physical interaction with the disassembly objects. It consists of a robot arm, a flipping table, and a grinder. The robot arm is used to move the grinder which is a disassembly tool and the flipping table is used to unload the detached components from the product (Table 6.1).

Table 6.1 Operating modules

Operating module	Associated hardware	Main functions	Language
CRM	Main computer	Cognitive robotic agent	IndiGolog
		Knowledge base	Prolog [7]
VSM	Depth camera	Detection functions	C/C++
	Colour camera		
DOM	Robot arm	Operation plan and primitive actions	RAPID [8]
	Flipping table		C/C++
	Grinder		

Communication System with a Client-Server Model

Internal communication is carried out in a straightforward manner within the scope of each operating module. A *data acquisition device* (DAQ) is needed in the case of the VSM and DOM since the main computer interacts with the hardware, namely the image grabber and I/O controller. The robot controller is comparable to the main computer and is capable of fully controlling the robot arm via the connected low-level hardware. An overview of the physical connections showing both internal and external communication is shown in Fig. 6.8.

The external communication among different modules is more complicated since the operating modules are developed on varying platforms. Therefore, the multi-platform client-server model [7] is selected to enable communication over the network. A communication protocol using Transmission Control Protocol and Internet Protocol (TCP/IP) based on the Prolog syntax is used to manage the information flow via socket-messaging. In this case, the messages are in the form of *actions* and *fluents* (see Chap. 5 for more detail). The protocol is designed to

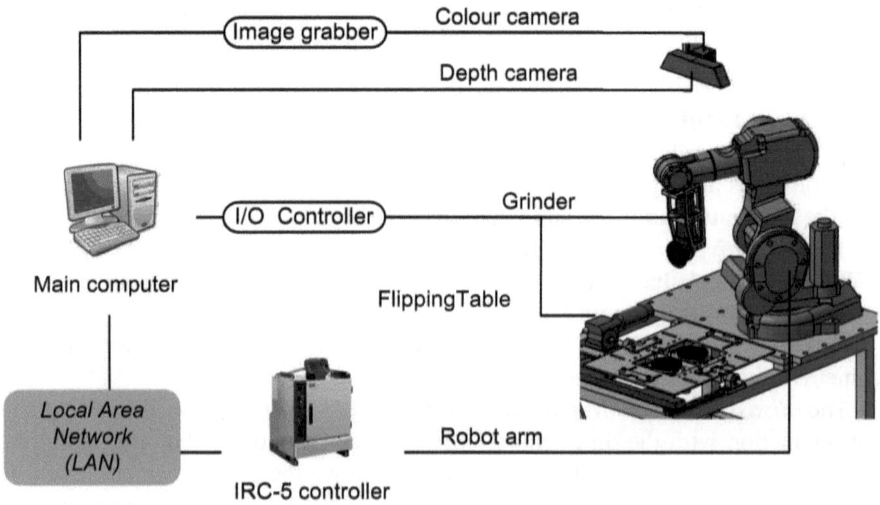

Fig. 6.8 Schematic diagram of the physical connections

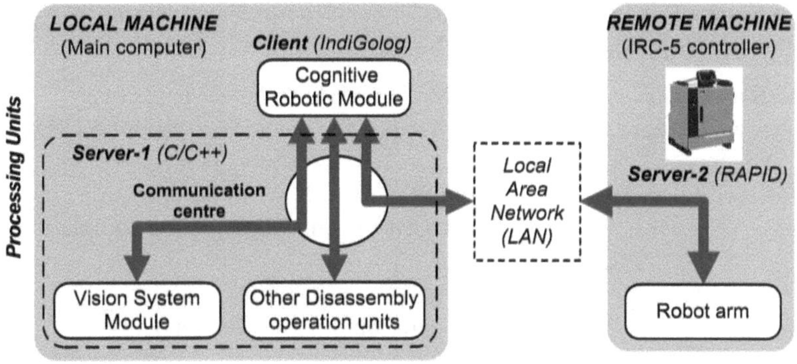

Fig. 6.9 Schematic diagram of communication network structure

be compatible with the cognitive robotic module, the planner which directs the operation of the other modules.

The network consists of three components: (a) *client*, (b) *server*, and (c) *communication centre*. In this case, the cognitive robotic module is a client and the other modules are servers (see Fig. 6.9). The client communicates with the servers by sending requests to the communication centre, either for sensing actions or primitive actions. Subsequently, the message is distributed to the corresponding module. Feedback is sent back from each module once the required information has been processed or the requested actions have been performed. The communication centre further adjusts the syntax as needed according to the different platforms. The communication system can be grouped, with respect to the location of the programs and language platform (see Fig. 6.9).

In Fig. 6.9, the client and Server-1 both operate on the local machine, located on the main computer. The client is the CRA which is the planner of the system. The TCP/IP connection is established via Prolog. C/C++ is used among many components on the local machine: the vision system, communication centre, and the disassembly operation module excluding the robot. These components are grouped together as Server-1 to simplify internal communication. Window Socket Library (winsock2) [8] is used to implement the communication protocol. Server-2 is the module located on a remote machine, the controller for the robot arm. TCP/IP for socket messaging is implemented in RAPID. This server connects to the communication network through the Local Area Network (LAN).

Interaction Between Modules—Actions and Fluents

With respect to the integration of the system, the actions and fluents associated with each module are listed. The actions are classified into three types: (a) primitive actions, (b) sensing actions, and (c) exogenous action, which are described in Sect. 5.3.3. Example of interactions between each pair of operating modules are given in Statements (5.6)–(6.3), where ‘s’ denotes a sensing action; ‘p’ a primitive action; ‘e’ an exogenous action; and, ‘f’ a fluent.

$$\begin{array}{ll} \textit{detectBackCover}. & \text{// s: CRM asks VSM to detect backcover} \\ \textit{box}(x_1, y_1, x_2, y_2, z_1, z_2). & \text{// f: VSM sends the detected location to CRM} \end{array} \tag{6.1}$$

$$\begin{array}{ll} \textit{cutRect}(x_1, y_1, x_2, y_2, z_1, m). & \text{// p: CRM asks DOM to do operation} \\ \textit{done}. & \text{// f: DOM informs CRM of the completion} \end{array} \tag{6.2}$$

$$\begin{array}{ll} \textit{senseHumanAssist}. & \text{// s: CRM asks human operator for assistance} \\ \langle \textit{demo action} \rangle. & \text{// e: human operation demonstrate an action} \end{array} \tag{6.3}$$

In conclusion, the system architecture is a composition of three operating modules that seamlessly connect with each other via the network system using a client-server model. The actions and fluents represent the commands and information used to interact among the modules. They are encoded in a form compatible with Prolog syntax, according to the client’s requirements. The messages are conveyed to the desired operating modules through the communication centre, which resolves the multi-platform compatibility problem. In the following sections, the operating modules are explained in detail. The design and the functional requirements according to LCD screens are described. The complete list of significant actions and fluents will be shown in the section of each module.

6.5 Cognitive Robotic Module

The cognitive robotic module is configured as presented in Chap. 5. In short, the CRM consist of CRA and KB. The CRA is formulated with the behaviour specification and the domain specification. The behaviour specification is influenced by the four cognitive functions and the domain specification is formulated with the disassembly domain. As this module controls the behaviour of the system, an overview of the entire disassembly process will be given in this section.

6.5.1 Design and Functions

Domain Specifications

The disassembly domain consists of three levels: (a) the product structure, (b) operation plan, and (c) process parameters (see Sect. 5.4.1). The state diagram of the product structure is constructed specifically for LCD screens (see Fig. 6.10). The information about this variation was gathered from around 40 different model samples. The products are disassembled into the module level. As described earlier in Sect. 6.2.3, two main structure types with six main components are considered

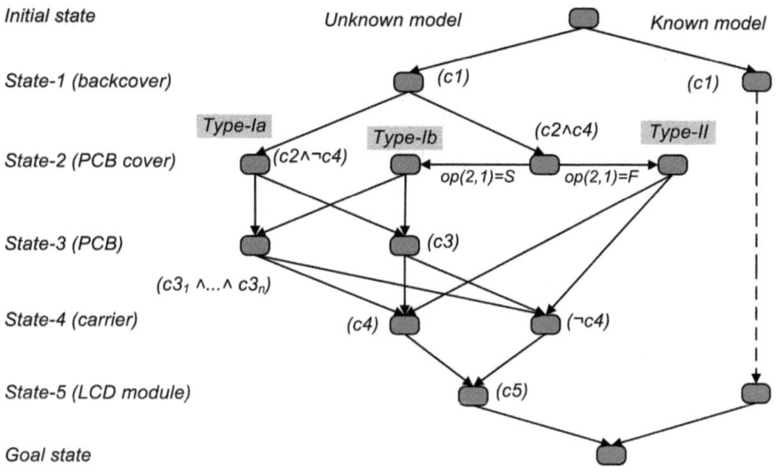

Fig. 6.10 Disassembly domain showing main structure of LCD screens

in developing this state diagram. Although significant variation in the ordering between PCB cover—PCBs—carrier is found, it is evident that the order of the main components is consistent in different models. The structure type can furthermore be identified based on the order of components detected. This is seen from State-2 to State-4 in Fig. 6.10 where both the visual detection outcome[2] and execution monitoring after performing a cutting operation[3] are needed to identify the correct state.

Basic Behaviour

In accordance with the (semi-)destructive approach, operation plans and parameters are designed based on the cutting operation using a grinder. Each type of component requires a different strategy for removal due to the varying types and locations of associated connective components. As a result, appropriate operation plans and process parameters need to be individually determined for each main component.

The operation plans are composed by the following parameterised primitive actions: (a) *cutContour*, (b) *cutCorners*, (c) *cutLine*, and (d) *cutPoint* (see complete list of actions in Appendix A). The inputs are fluents representing primitive geometries (i.e. *rect*, *line*, and *loc*) according to each primitive action. The initial investigation suggested that due to the simple structure of LCD screens, it is

[2] In Fig. 6.10, notation in the bracket is the combination of the components detected or not detected at that state, e.g. ($c_i \wedge c_j$) denotes the state that c_i is detected whereas c_j must not exists.

[3] The executed *j*th-operation plan of component c_i is denoted with *op*(*ci,j*). 'S' and 'F' represent the execution result, success and failure, of removal the component.

sufficient to define all cutting actions on horizontal planes. The cutting location on this horizontal plane is parameterised with respect to the sensed component location, fixed via an offset according to the operation plan. The required depth of cut is initially unknown, and incrementally changes in each operating cycle until reaching a critical depth which results in main component removal. The feed speed and the orientation of the cutting tool are variable and considered process parameters. In short, 1–4 operation plans (see Eq. (6.4)) with 3 process parameters (see Eqs. (6.5) and (6.6)[4]) are used depending on the type of component. The set of possible variations can be seen from these definitions (as detailed in Sect. 6.7.2).

$$operation\ plan = op(ci,j) | i \in \{1, 2, \ldots 6\} \wedge j \in \{0, 1, \ldots, 3\} \tag{6.4}$$

$$m \in \left\{\theta_{tool} \times S_{feed}\right\} = \{\text{'0'}, \text{'1'}, \ldots \text{'8'}\} \tag{6.5}$$

$$z \in \left\{z_{F\min} \leq z_F \leq z_{F\max} | z_F \in \mathbb{Z}\right\} \tag{6.6}$$

$$\Delta z \in \left\{\Delta z_{\min} \leq \Delta z \leq \Delta z_{\max} | \Delta z \in \mathbb{Z}\right\} \tag{6.7}$$

Execution monitoring is implemented at the planning and the operation in a specific way. At the *planning level*, the removal of a main component is assessed by visually detecting the state change. The sensing action *detectStateChange* is executed after executing each cutting operation plan, followed by an attempt to remove the component. In this case, the primitive action *flipTable* flips the fixture holding the product in place, causing the component to fall and be collected if the threshold depth has been reached (see Sect. 6.7.2). Execution monitoring is used to classify the type of main structure as shown in State-2 of Fig. 6.10. The success or failure of the operation op(*c2,1*) differentiates between Type-Ib and Type-II.

The success of a cutting operation without collision is also an important source of information in the *operation level*. A successful completion of the cutting path implies that cutting tool is able to reach the cutting destination using the current cutting method *m*. The robot is hence capable of determining a suitable cutting method by itself. The successful cutting method is acknowledged by sensing action *checkCuttingMethod*. Subsequently, since the destination can be vertically approached, this *m* can be used in the next operation cycle to repeat the horizontal

[4] The cutting method *m* represents a combination of tool orientation θ_{tool} and feed speed S_{feed}. In this case, the combinations are denoted with '0'…'8'. The z_F denotes the level relative to the fixture and Δz is incremental step changed. z_F and Δz are discrete value as they are integers.

cutting path with an adjusted depth. In summary, the operation routine for operation plan *op(ci,j)* with execution monitoring for this system is in Procedure (6.8)

$$
\begin{array}{ll}
\textbf{Proc } op(c_i,j) & \\
\quad \textbf{while } (\neg stateChange \wedge m \neq \text{'0'} \wedge (z_{Fmin} \leq z_F \leq z_{Fmax})) & \\
\quad\quad offsetPrimitiveDepth & \text{// incrementally change } z_F \\
\quad\quad cutPrimitive & \text{// exeute cutting operation} \\
\quad\quad checkCuttingMethod & \text{// obtain fluent cuttingMethod} \\
\quad\quad flipTable & \text{// remove the detached part} \\
\quad\quad detectStateChange & \text{// obtain fluent stateChange} \\
\quad \textbf{endWhile} & \\
\textbf{endProc} &
\end{array}
\tag{6.8}
$$

Advanced Behaviour

The knowledge base, learning, and revision are implemented as explained in Sect. 5.5 but applied more specifically to LCD screens. Since the operation plans are pre-defined, the number of the operation plans is known a priori. Therefore, the agent's program is simplified by limiting the learning to this known structure. According to Expression (5.7), the Φ_i store the primitive geometry of the horizontal path of the desired cutting destination. Γ_i are the parameters for the tool's approach. The first part, representing the operation for connection disestablishment, refers only to screw removal.

As a result, a simplified version of the general plan is given in Eq. (5.8), where the maximum number of operation plans available for the main components is four. Γ_0 for screw cutting represents only the cutting method since the cutting depth for the screw head is quite consistent. On the other hand, because the thickness of components is unknown, the other Γ_i must be additionally specified with a starting depth z_{start} and end depth z_{dst} in additional to m.

$$plan_{general} = \left[\left[\vec{\Phi}_0\right], \left[\vec{\Gamma}_0\right], \Phi_1, \Gamma_1, \Phi_2, \Gamma_2, \Phi_3, \Gamma_3\right] \tag{6.9}$$

where

$$
\begin{aligned}
\left[\vec{\Phi}_0\right] &= \left[loc(x_{10}, y_{10}, z_{10}), loc(x_{20}, y_{20}, z_{20}), \ldots loc(x_{n0}, y_{n0}, z_{n0})\right] \\
\left[\vec{\Gamma}_0\right] &= [m_{10}, m_{20}, \ldots m_{n0}] \\
\Phi_1 &= rect(x_{11}, y_{11}, x_{21}, y_{21}, z_{dst1}) \quad \text{and} \quad \Gamma_1 = [m_{11}, z_{start1}] \\
\Phi_2 &= line(x_{12}, y_{12}, x_{22}, y_{22}, z_{dst2}) \quad \text{and} \quad \Gamma_1 = [m_{12}, z_{start2}] \\
\Phi_1 &= rect(x_{13}, y_{13}, x_{23}, y_{23}, z_{dst3}) \quad \text{and} \quad \Gamma_1 = [m_{13}, z_{start3}]
\end{aligned}
$$

Regarding new plans contributed via assistance, demonstrated primitive actions are encoded as shown in Eq. (6.10). In the GUI console, the operator demonstrates an action by first selecting it from the provided options then drawing the

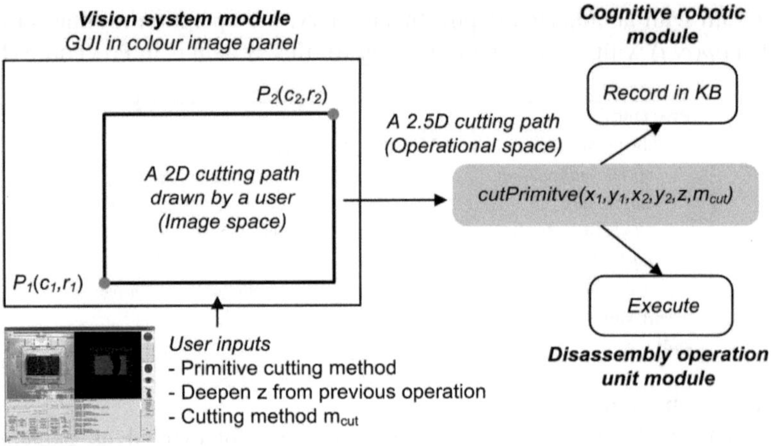

Fig. 6.11 User's demonstrated primitive cutting operation in GUI

cutting path on the 2D colour image in the GUI console (see Appendix B). This specified cutting path is then converted to a 3D path with the aid of the vision system (see Fig. 6.11).

$$plan_{add-on} = [cutContour(x_{11}, y_{11}, x_{21}, y_{21}, z_{dst1}, m_{11}), \ldots] \qquad (6.10)$$

Other exogenous actions (e.g. *skipComponent*, *newComponentLocation*, etc) allow the operator to resolve inaccuracies in the system's original belief. These actions implicitly change the corresponding fact in the planning level KB and do not alter facts in the operation level.

Once a model has been learned in the KB, regardless of applied revision strategy, faster performance is achieved by reducing the number of operation cycles using larger cutting steps. Since the final cutting destination is known, the agent can simply approach this desired depth without frequently assessing the state change (a time consuming process).[5] Significant time is reduced by assessing state change only once, after the final operation cycle at end of the operation plan.

For revision, the simple approach explained in Sect. 5.5.4 is implemented. The order of application of the operation plans is designed in regard to the impact on the components and success removal rate. The cutting screw operation is first executed since it is expected to cause minimal damage to the component. Afterwards, higher impact operations corresponding to different offsets from the edge of the component are used. The result from this revision strategy will be clearly shown in Sect. 6.8.4.

[5] To assess state change, the action *flipTable* is executed in addition to visual detection. This physical operation takes around 8.5 s.

6.6 Vision System Module

The VSM addresses the uncertainties associated with the quality and quantity of the components. In particular, the vision system is designed to handle variations in (a) physical appearance in the components, (b) quantity of the component, and (c) location of the components. As the system is not aimed for component reuse, it is sufficient and more robust to simply classify components by type. However, components of the same type typically appear different in each model, even within the same product family. The system must be capable of making the classifications despite of these variations. The CRA requests information using a sensing action. Subsequently, the vision system attempts to recognise the specified component and determines the number and location of each component detected. The uncertainties are suppressed by processing the raw detection outcome into abstract information. This information is sent back to the CRA via fluents for further decision-making. Sensing actions and associated fluents are listed in Appendix A.

6.6.1 Design and Functions

This module needs to perform the detection of the main and the connected components. The detection process can be described as two processes, (a) recognition and (b) localisation, which are used to address all relevant uncertainties. Moreover, the vision system is used to determine a transition of the disassembly state, facilitating the cognitive robotics task of execution monitoring.

Hardware Setup and Coordinate Mapping
The disassembly process is performed on the product in 3D space. Two cameras, i.e. a colour camera and a depth camera, are used to obtain this information (see detail in Sect. 4.2.3). This depth camera also satisfies the low-cost constraint in contrast to the other 3D imaging technologies.

These two cameras are mounted stationary above the flipping table where the whole LCD screen can be seen from the back side. According to this setup, calibration is needed in three areas (a) colour, (b) image quality, and (c) coordinate mapping. All calibration methods are implemented in the low-level control layer. First, accurate colour perception is necessary since colour-based recognition techniques are used for some components. In this case, colour balance correction is performed [9]. The second issue, regarding the image quality, deals with noise and inaccuracy of sensed information. This problem is greater in the depth image, which is sensitive to edges and highly reflective surfaces and results in inaccurate data sensed from these areas. In addition, due to the depth resolution of approximately 3–4 mm, position error due to noise is significant. Therefore, noise filtering and interpolation techniques are needed to improve image quality.

Finally, coordinate mapping is crucial for the entire process due to the requirement for localisation of the component and positioning of the related cutting operations. Therefore, the relation between the product coordinate frame {P} and the image space {S} need to be determined. In this case, the calibration process can be simplified by mechanical adjustment. The two cameras are aligned parallel to each other such that the lines of sight are normal to the fixture plate. As a result, the number of variables reduces significantly. Eventually, Eq. (4.11) is used for localisation, where position in the operational space (x_p, y_p, z_p) of the object relative to {P} is a function of the image space variables (c, r, z_F). The c and r are obtained from the colour camera and z_F from the depth camera (see detailed explanation is in Sect. 4.2.4).

$$\mathbf{P}^{P}_{object}(x_P, y_P, z_P) = \mathbf{H}(c, r, z_F) \quad (6.11)$$

Detection of Components

According to the structure analysis of LCD screens and the actual disassembly process, 5 main components and 1 connective component need to be detected: (a) back cover, (b) PCB cover, (c) PCBs, (d) carrier, (e) LCD module, and (f) screws.[6]

The component types are recognised using rules based on *common features* as summarised in Fig. 6.4 (see detail of common features in Sect. 4.4.2). The information from both types of images is taken into account. In this case, the common features are determined from the 41 initial samples manufactured between years 1999 and 2011. The presented values are subject to change when applied to newer models (Table 6.2).

The *back cover* is the outermost part made of plastic that covers all other components within the product. The size and aspect ratio vary according to the size of the LCD panel, and in this case, is limited to 15″–19″ with a maximum aspect ratio of 2:1. The height of the back cover can be used to reject other small irrelevant objects, e.g. supporting fixture elements. From the samples, the thickness is 10–70 mm from the front of LCD screen. However, for newer design the LCD screen tends to be thinner (see detection example in Fig. 6.13a).

PCB cover is a metal box that covers PCBs. The height of the box used to identify this component. This component can be a separate component (Type-I) or integrated as part of the carrier (Type-II). The classification between subclasses can be preliminarily done by colour criteria. The Type-Ia cover is typically made of reflective sheet metal, approximately 0.5–1 mm thick. Type-Ib and Type-II covers typically consist of a matte gray metal that is 1–3 mm thick due to the strength required for the carrier. The colour range for the reflective metal is $H \in (35°, 130°)$ and $S \in (12, 35)$. The colour range for the matte gray metal is $H \in (73°, 135°)$ and $S \in (10, 27)$ (see detection example in Fig. 6.13c).

PCBs are usually perfectly-rectangular plates found in two colours: green and yellow. The colour range criteria are, respectively, $H \in (70°, 200°)$ and $S \in (35, 80)$ for green, and $H \in (20°, 70°)$ and $S \in (35, 90)$ for yellow. Due to a number of

[6] Note: the front cover is excluded from the list since it is expected to be detached according to operations in earlier states. Other types of connections, i.e. cables, snap-fits, and plastic rivet, are also expected to be disestablished by the operation plan. As a result, visual detection of these components is unnecessary.

Table 6.2 Common features for the detection of the components in LCD screens

Component	Component type (m—main and c—connective)	Recognition	Localisation	Common features							
				Colour image						Depth	
										Image	
				2D geometry			Colour range	Connected region (blob)	Haar-like feature	Height	Surface roughness
				Size	Shape	Aspect ratio					
Back cover	m	•	•	•	•	•				•	
PCB cover	m	•	•	•	•		•	•		•	
PCBs	m	•	•	•	•		•	•			
Carrier	m	•	•		•		•	•			
LCD module	m	•					•	•		•	•
Screws	c	•	•	•					•		

embedded components mounted on the boards, the number of the connected pixels belonging to an area having valid a colour is taken into account to accurately determine the boundary of this component. Blob detection [10] is used to detect regions in which the area of the bounding rectangle lies between 1,000–40,000 mm^2. In addition, since multiple PCBs may be found in one disassembly state, an algorithm further separates the detected area into isolated PCBs (see detection example in Fig. 6.13e).

The *carrier* is a core structural component of LCD screens, to which other main components are mounted. It is designed to be high strength; a thick matte gray metal is always used. The colour criterion is used to identify this. The size of this component is large enough to support the LCD module. It is determined by a connected region of the specified colour, covering 45–90% of the LCD screen (see detection example in Fig. 6.13g).

The *LCD module* is the component that performs the main functionality of the product, and appears as a perfect rectangle covering more than 90 % of the entire area of the product. The back of LCD module is flat and usually consists of shiny sheet metal which can be identified using a colour criterion and/or surface roughness calculated from the depth image. The LCD module is then detected by determining the appropriate connected region satisfying the surface criteria. Detection of the LCD module is important because after this is achieved the CRA will acknowledge that the goal state has been reached (see detection example in Fig. 6.13i).

Screws are the only connective component to be detected in this system. In comparison with other fastener types, screws are rigid and well-defined, simplifying the visual detection task. In addition, most other connectors are expected to be automatically disestablished by the cutting operations. From the top view, the heads of M2–M3 flat button Phillips screws are detected using a classifier based on Haar-like features [11]. The classifier has been trained by approximately 800 positive images of screw's head and 7,000 negative images. A high detection rate is achieved, however alongside a significant number of false negatives; the size criterion is used in order to filter out some false negative detected objects. The location of the attached main components is also taken into consideration in order to limit the area to be searched. For example, in a process of removing a PCB, the region of interest (ROI) will be set to cover the PCB and the screw detection will be performed only in this area.

To communicate abstract localisation information, a main component is represented by a minimal bounding rectangle (MBR) and a screw by a point in 3D operational space. The fluents are $compLocation = box(x_1, y_1, x_2, y_2, z_1, z_2)$ and $screwLocation = loc(x, y, z)$, respectively. The position in the operation space is obtained from Eq. (4.11). In case that multiple instances of a type of component is detected, the locations are represented in a list, e.g. $pcbLocation = [box(x_{11}, y_{11}, x_{21}, y_{21}, z_{11}, z_{21}), box(x_{12}, y_{12}, x_{22}, y_{22}, z_{12}, z_{22}), \ldots box(x_{1n}, yn_{1n}, x_{2n}, y_{2n}, z_{1n}, z_{2n})]$.

Detection of State Change and Other Utilities

The detection of state change is performed according to the algorithm presented in Sect. 4.4.4. The change of disassembly state is measured from the differences and similarities between the original condition and the current condition of a disassembly state. The colour and the depth images are considered over the ROI. However, the depth image is more critical, as it represents the physical geometry of the component. The state change is assessed using Eq. (4.23). A depth difference threshold (Φ_{depth}) of 50 % and a colour difference threshold (Φ_{colour}) of 75 % are selected. Experimentally, this was found to successfully recognise state change in 95 % of the samples.

Other functions of the vision system include: (a) recognition of the product model and (b) height measurement z_F of features.

The product model detector is used to determine whether the current sample constitutes a previously-seen (known) model. This recognition is performed using SURF [12] on the colour image in order to match the keypoints of the current model to the existing database. It was experimentally found that a match percentage of greater than 15 % implies that two samples are identical.

The measured height z_F is used as the starting point of cutting operation for the known models (see Fig. 6.12), and ensures that the cutting operation starts from the top surface, preventing force overload on the cutting tool. Force overloading cut can occur when the cutting tool tries to quickly approach the target by cutting through thick layers of materials. This inappropriate movement will cause the cutting tool to be blunt, worn, or broken. Step cutting shown in Fig. 6.12 is one of the strategies to prevent this circumstance (Fig. 6.13).

6.6.2 Performance

The detectors were tested on 37 different models to validate the selected rules and common features. The performance in the recognition and localisation processes based on the non-damaged components is shown in Table 6.3. The recognition of

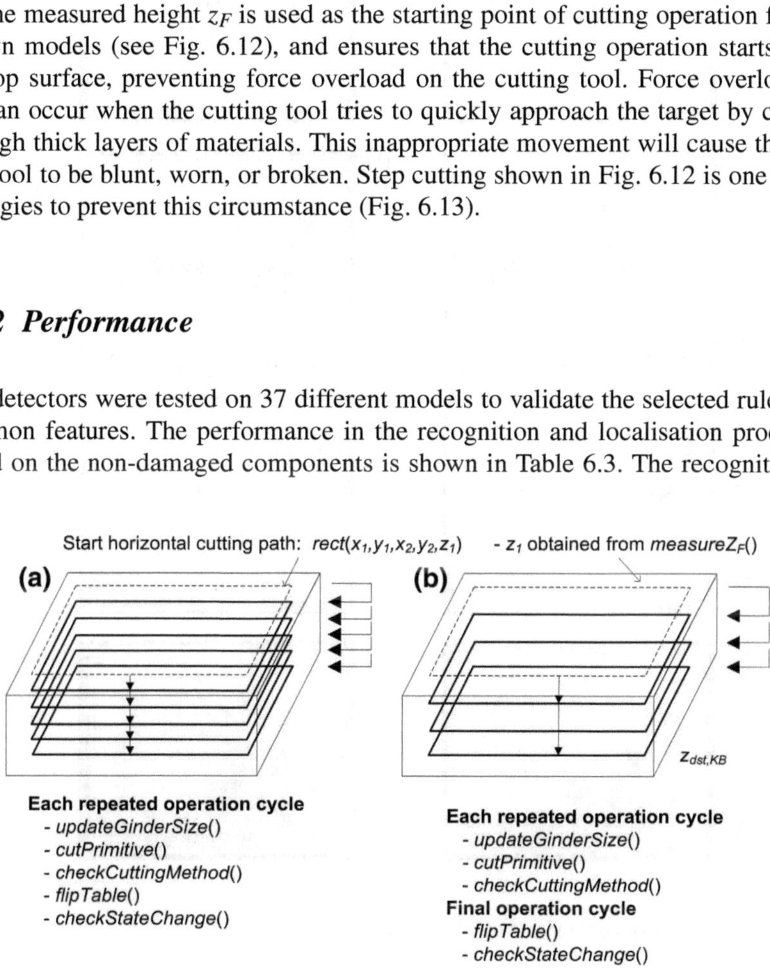

Fig. 6.12 Incremental cutting operations. **a** Unknown model. **b** Known model

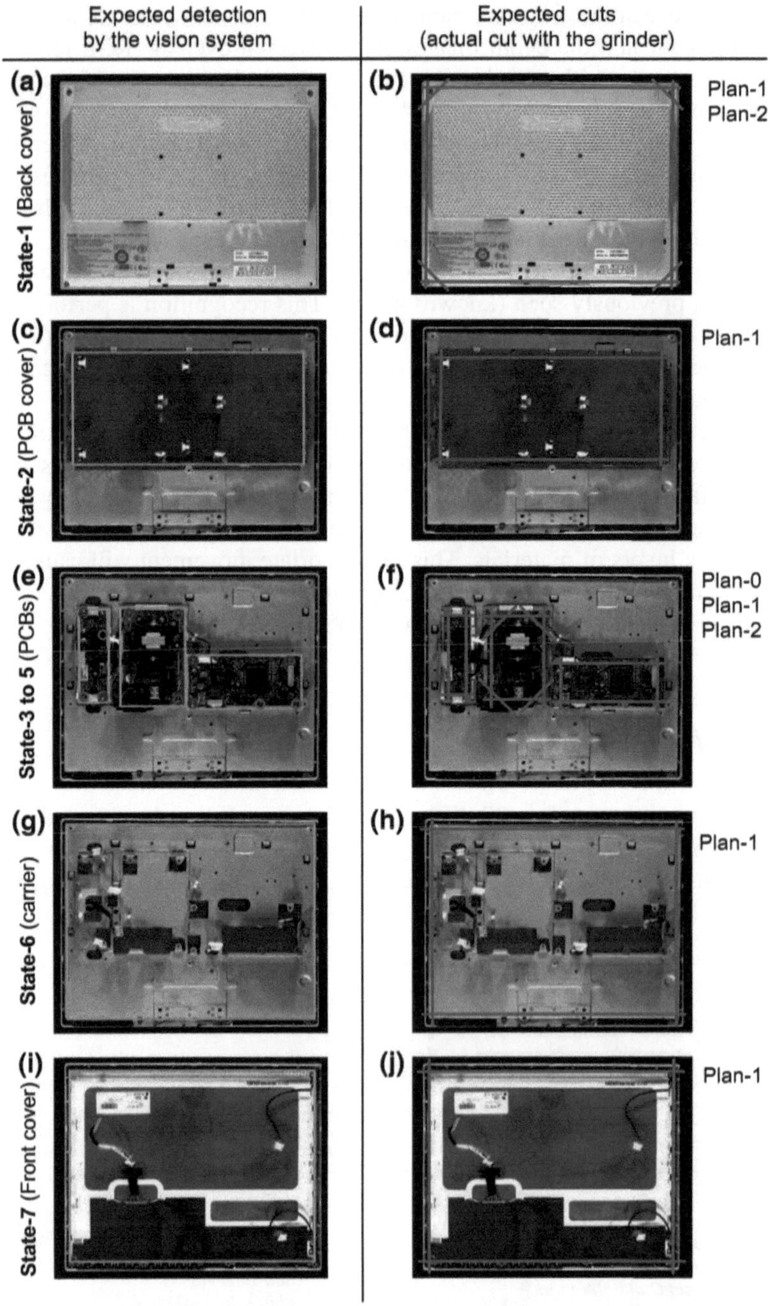

Fig. 6.13 Samples of components in LCD screens

Table 6.3 Summary of operation plans for removing the main components

Main component	Operation plan			
	Plan No.	Primitive cutting operation	Inner offset from the border (mm)	Intention or connection to be disestablished
Back cover (c1)	1	cutContour	5	Snap-fits around border area
	2	cutCorner	20	Screws at the corners
	3	cutContour	12	Screws around border
	H	*	*	– Press-fits at with the PCB cover – Uncut screws in the middle area – Correct the inaccuracy
PCB cover (c2)	1	cutContour	5–10	Remove the top plate
	2	cutContour	5 (outer)	Hidden cables underneath the carrier
	H	*	*	– Uncut hidden cables – Correct the inaccuracy
PCB (c3)	0	cutScrew	n/a	Screws at particular locations
	1	cutContour	5	External ports, cables, screws
	2	cutCorner	20	Screws at the corners
	3	cutContour	10	Screws around border area
	H	*	*	– Uncut connectors in the middle area – Correct the inaccuracy
Carrier (c4)	1	cutContour	5	Screws, snap-fits around border area
	H	*	*	– Correct the inaccuracy
Front cover (c6)	1	cutContour	5 (outer)	Snap-fits around border area
	H	*	*	– Correct the inaccuracy

Note * Based on the decision of the operator; *H* human assistance; Operation plan used in cognitive robotics denotes as op(⟨MainComponent⟩,⟨plan No.⟩), e.g. op(c1,1) represents operation plan-1 for treating a back cover

components generally performed with around 80 % accuracy, with an exception of screws and the LCD module, which had an accuracy of 65 and 50 % respectively. Regarding localisation precision, the position of detected components was determined with a Root Mean Square (RMS) error of within 6 mm. State change can be assessed with 95 % accuracy. In general, the error of the detection algorithms results from the depth image due to the Kinect sensor's sensitivity to reflective surfaces and edges. In addition, the colour range used to determine connected regions can be slightly inaccurate due to the reflection of colour from other nearby components.

In conclusion, the VSM is capable of detecting the components effectively. Errors resulting from the uncertainties in the component and non-detectable objects are expected to be further resolved by the operation plans and assistance.

6.7 Disassembly Operation Unit Module

The DOM addresses the problem in the disassembly operation due to the physical uncertainties of (a) EOL condition and (b) other non-detectable objects. These can be considered as missing pieces of information, which the sensing module alone is incapable of obtaining. Variations in EOL condition manifests itself as minor physical changes in the product returned, e.g. in the case of damaged parts. Problems caused by non-detectable objects occur in two primary categories: undetected connected components, and structures obstructing robot movement during the disassembly operation. The CRA can only control the process appropriately if such information can be perceived. In order to fulfil this functional requirement, the disassembly operation units need to appropriately respond to these uncertainties using hardware and return relevant information to the CRA.

In the case study, the system is design to be simultaneously low-cost, flexible and robust to uncertainties. Therefore, the disassembly rig, including a robot and disassembly tools, are designed to perform (semi-)destructive disassembly. In addition, operation plans based on common connection locations is used, designed to have a high chance of applicability and success even on unseen models.

6.7.1 Hardware Design and Functions

As earlier described in Sect. 3.5, the disassembly rig consists of three operation units: (1) a robot arm, (2) a flipping table, and (3) an angle grinder. The robot arm is a stand-alone module, used as the main element of the DOM. The main task is to control the disassembly tool to perform the (semi-)destructive disassembly operations according to the primitive action requested by the CRA. The flipping table is used as a device for handling an LCD screen while the grinder is used as a disassembly tool (see Fig. 3.17). The CRA can simply activate these devices with the primitive actions *flipTable* and *switchGrinderOn/Off*.

The control architecture is more complicated or the robot arm module. In the *low-level control* layer, the robot controller controls the motion of the robot arm at the motor-sensor level to an accuracy of within 1.0 mm. The part directing the operation is in the *mid-level control layer*. The parameterised operation procedures that correspond to the primitive actions are programmed in this part. The operation procedures are *cutLine, cutContour, cutCorner,* and *cutScrew*. The supplied parameters are the cutting location and cutting method (i.e. feed speed and tool orientation).

The robot controller also sends feedback to the CRA, informing it of the parameters leading to successful execution. The operation succeeds if the entire cutting path can be cut without any collisions. At the end of each operation, the CRA requests *senseCuttingMethod*, at which the robot controller returns the cutting method m. This was earlier presented in Eq. (6.5). m is defined in

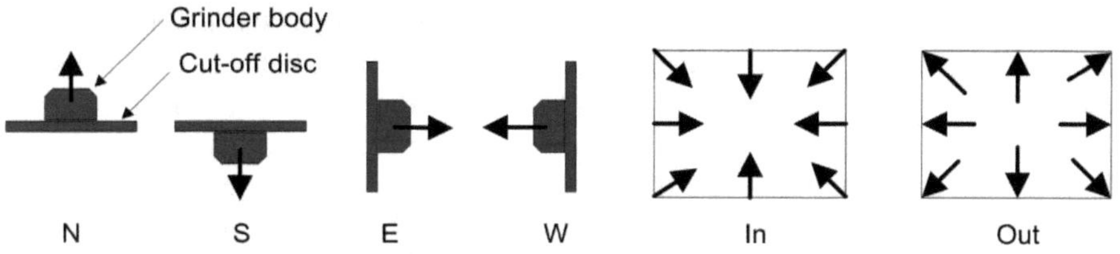

Fig. 6.14 Notation of tool's orientation seen from top view

Eq. (6.12). $M_{success}$ is encoded with the values '1'–'8' representing the tool orientation and feed speed used in the successful operation, whereas M_{fail} is represented with '0'. The successful cutting method is subsequently recorded in the learning process. The notation of tool orientation is shown in Fig. 6.14. The orientations {N, S, W, E} are used with *cutLine*, whereas only {In, Out} are used with *cutContour* and *cutCorners*.

$$m = \begin{cases} M_{success} \in \theta_{tool} \times S_{feel} \big| (\theta_{tool} \in \{N, S, W, E\} \vee \theta_{tool} \in \{In, Out\}) \wedge S_{feel} \in \{L, H\} \\ M_{fail} \end{cases} \tag{6.12}$$

6.7.2 Functions

This section presents the *general operation plans*, procedures in the CRA containing sequences of parameterised primitive actions applicable to general (and previously unseen) models of LCD screens. The operation plans are designed based on strategies to disestablish the connections of each main component. In the treatment of a main component, the available operation plans are called one by one in an order specified by the CRA. In formulating the operation plans, three main factors are taken into account: (a) *common location of the connectors*, (b) *level of destruction*, and (c) *success rate of the removal process*.

The common location of fasteners is specific to each type of component. In general, connectors typically lie nearby the border of each main component. Only a small amount is found in other locations. The general operation plans are used in the former case and human demonstration in the latter, respectively. For the general plan, the *border offset* is a significant factor in determining the cutting location. Appropriate parameters can be obtained statistically, from observing the common locations of connectors from samples of different models [13].

Regarding the level of destruction, *semi-destructive disassembly* is preferred, whereby only the connective components are damaged without or minimally damaging the main component. However, the success rate is lower since due to the possibility of misdetection of the connective components. For the *destructive disassembly*, the cutting operation is performed directly on the main components. Both the success rate and the level of destruction are higher in this case. In contrast to semi-destructive disassembly, it is possible here to disestablish unseen connectors that lie underneath the component, and separate the main fraction of a component while leaving inseparable parts behind. This latter is the case when the majority of the back cover is detached by cutting it away from the snap-fits located in the border. The size of the successfully detached part can be maximised by selecting the minimal border offset that provides an acceptable success rate.

The common locations of the connectors for each type of component, and the corresponding operation plans, are explained as follows. The connections and disassembly processes are analysed in the perspective of the view and operations available to the system. Therefore, the potentially-erroneous outcome from the vision system is taken into account.[7]

Back Cover

The back cover is generally connected to the front cover or carrier with 8–10 snap-fits within 5 mm around the border area. In addition, 4 screws are commonly found close to the corners within 10–12 mm of the border, and 1–2 additional screws are occasionally found in the central area for additional support. In certain cases, hidden press-fits also occur in the central area supporting the PCB cover.

The operation plans consist of cutting at different offsets from the border using *cutContour*, then cutting at the corners to bypass potential corner screws. The operations cut into the component to a limited depth with the aim of detaching the majority of the material from the product. The common location of the connections and the respective operation plans are shown in Fig. 6.15.

PCB Cover

For Type-I LCD screens, the PCB cover is generally a box that attaches to the carrier with snap-fits and screws around the base (see Fig. 6.16a). These connectors are not needed in Type-II structures where the cover is part of the carrier (see Fig. 6.16b). In both cases, there are cables and external ports joining parts of the PCBs that lie inside the cover. These cables and ports sometimes need to be disestablished in order to remove the PCB cover.

For the operation plan, *cutContour* with different offsets is used to cut the top surface and at the base level. This operation is needed to distinguish between two structure types. Plan-1 ($op(c2,1)$) cuts through the top surface of the PCB cover. In a Type-I structure, the top section of the cover is now detached, since there are no remaining connections (see Fig. 6.16a). In contrast, for Type-II, the cover remains connected to the rest of the product via the PCBs and the external ports (see Fig. 6.16b). As a result, the product structure type can be identified using the combination of an action and execution monitoring (see basic behaviour in Sect. 6.5.1). For Plan-2, the cut is done at the base level in order to disestablish the connections including the hidden cable underneath the carrier. The common location of the connections and the operation plans are shown in Fig. 6.17.

[7] Semi-destructive disassembly for removing screws is not used for most main components due to the low screw detection accuracy. Destruction of screws is only considered in the case of PCBs where the screw detection rate is high.

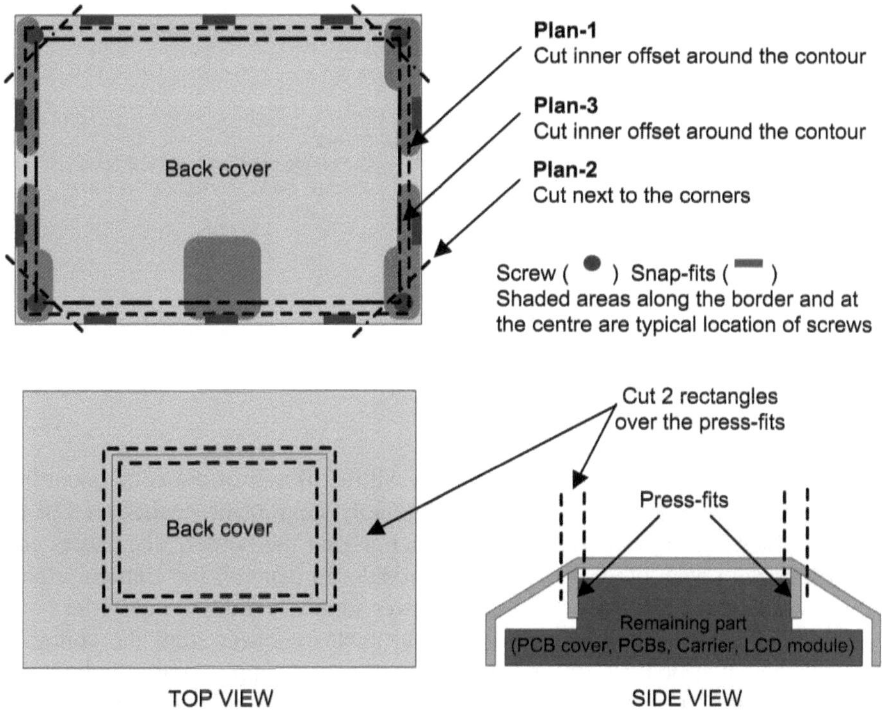

Fig. 6.15 Operation plans of the back cover

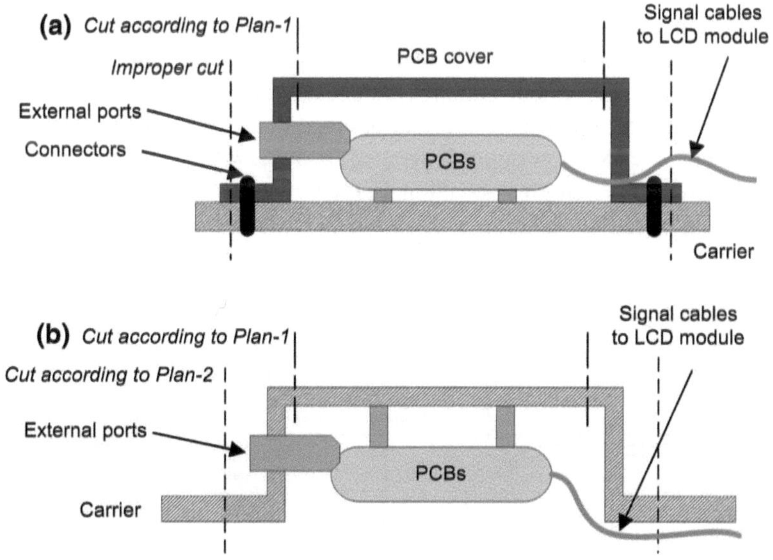

Fig. 6.16 Cross section of PCB covers in the different structure types. **a** Type-I. **b** Type-II

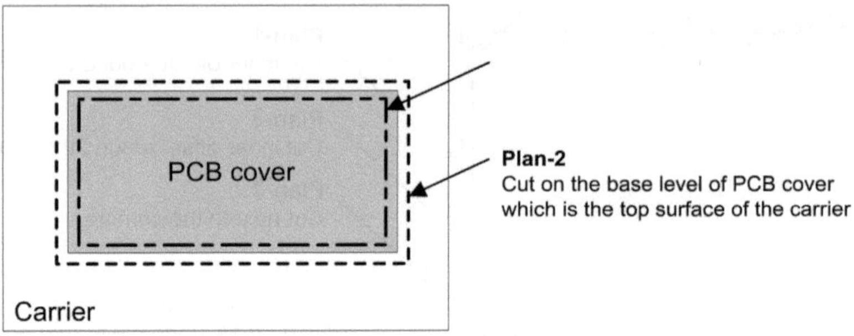

Fig. 6.17 Operation plan for PCB cover

PCBs

PCBs connect to the carrier via 4–5 screws within 10 mm of the component border. In some cases, 1–2 plastic rivets additionally support the central area of the board. Cables and input ports are also found around the border. The cables connect PCBs with each other and the LCD module. In general, the configuration is as follows: Power cables connect from a power supply board to a controller board and the CCFLs in the LCD module. Signal cables connect from the controller board to a switch board at the front panel, the power supply board, and another PCB that is part of the LCD module. The cables connecting to the LCD module are normally hidden underneath the carrier. The configuration of PCBs is different in Type-I and Type-II as shown in Fig. 6.18.

For the operation plan, both semi-destructive and destructive approaches are applied to this component. To minimise damage to the PCB, *cutScrew* is performed first, the screws on this component are easily detected and this operation causes minimal damage to the PCB. Afterwards, *cutContour* and *cutCorners* with various offsets are performed. The first cut occurs very close to the border in order to cut the cables and ports with minimal damage to the PCB. The second cuts the corners where screws are commonly located. Lastly, *cutContour* with a larger offset is performed. The common location of the connection and the operation plans are shown in Fig. 6.19.

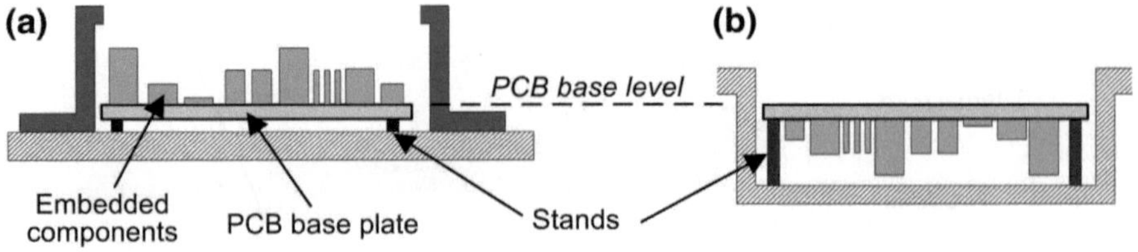

Fig. 6.18 Configuration of PCBs. **a** Type-I. **b** Type-II

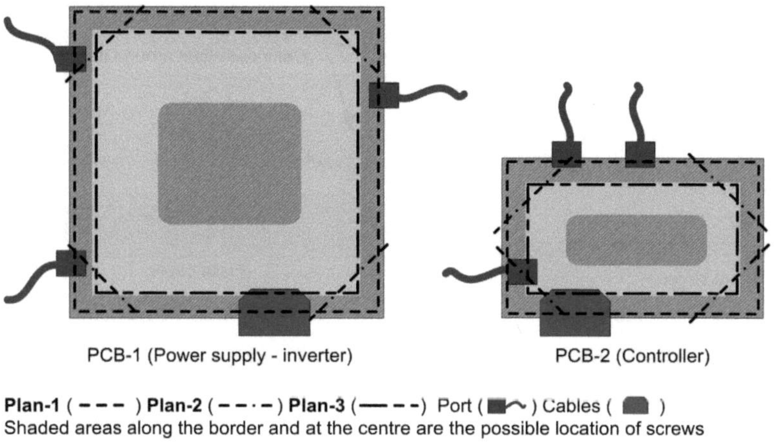

Fig. 6.19 Operation plans of the PCBs

Carrier

The carrier connects the PCBs to the PCB cover, and attaches to the LCD module via 6–8 snap-fits and 4–8 screws around the border. In most cases, the cables from the PCB are attached to the carrier via clips and screws, which also need to be disestablished. Furthermore, in the Type-II structure, where the carrier also covers the PCBs, input ports located around the covering area also need to be detached. However, this should have already occurred during removal of the PCB cover.

Regarding the operation plan, *cutContour* is performed, cutting near the border of the carrier to detach the majority of the component material. One of the challenges of this operation is to identify an appropriate depth of cut that avoids damaging the LCD module underneath. In this case, the preferred depth of cut is 2–3 mm measured from the carrier's top surface. The common location of the connection and the operation plans are shown in Fig. 6.20.

LCD Module and Front Cover

The LCD module is typically connected to the carrier, as well as to the front cover via snap-fits around the border. At this state of the disassembly process, all connections are expected to have been disestablished by earlier operations. However, in case that the LCD module is still attached to the front cover, an extra operation can be done to cut around the contour of the front cover, destroying all remaining connections. This is the final state of the disassembly process. At this stage, the LCD module can finally be separated.

In conclusion, general operation plans are used to disestablish connections in the treatment process of each main component. The plans are summarised in Table 6.4. The CRA will request each operation using primitive actions. The general plans are designed to be flexible and effective in detaching the main component, typically via a destructive approach. The strategies developed are

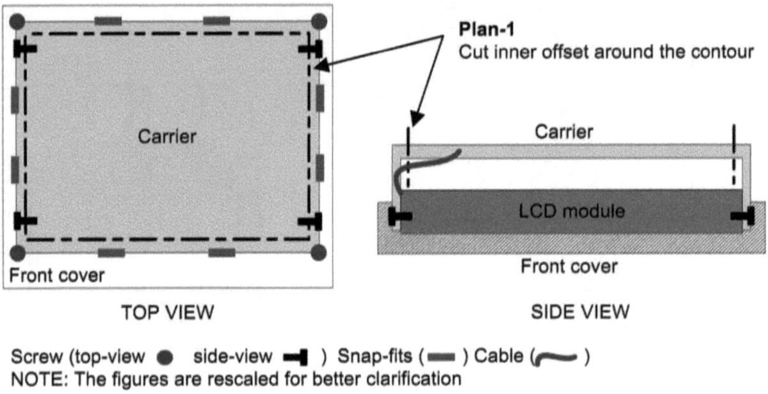

Fig. 6.20 Common location of the connections belonging to the carrier

Table 6.4 Success rate of the plans for removing main components

Main component	Rate of success to remove the main component (%)				
	General plans conducted autonomously				Human assistance
	Plan-0	Plan-1	Plan-2	Plan-3	
Back cover	n/a	0	0	12.5	100.00
PCB cover	n/a	46.67	33.33	n/a	100.00
PCB	0	0	0	17.65	100.00
Carrier	n/a	16.67	n/a	n/a	100.00
LCD module	n/a				

based on the common location of the components and corresponding connections as sensed by the vision system. Hence, the parameters, i.e. border offset values, implicitly account for errors in visual localisation. The uncertainties in regard to visually non-detectable connections and inaccuracy detection are also addressed.

6.8 Experiment

6.8.1 Process Flow

This section aims to demonstrate the operation of the system by describing the process flow relating to initial disassembly, learning and revision of the KB. The descriptions focus on key situations that emphasise the characteristics of the cognitive robotics approach. An example of the KB in regard to learning and revision process is shown, as well as the visual inputs and demonstrated actions. Images showing a typical disassembly operation are shown in Fig. 6.21.

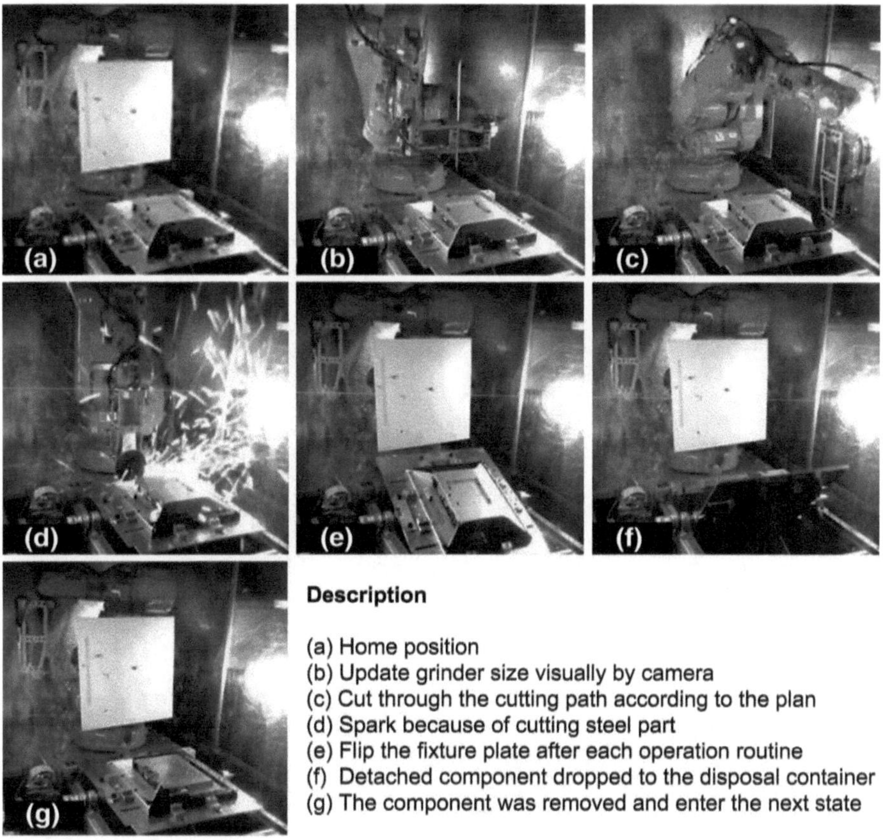

Fig. 6.21 Snapshots of the disassembly process captured from the test

Type-I Versus Type-II

Due to the variations in product structure, the disassembly process varies between Type-I and Type-II LCD screens. As shown in Fig. 6.10, a major difference occurs at the states representing the removal of PCB cover, PCBs and carrier. As a result, Type-I samples can be completely disassembled in a single run, whereas Type-II samples generally need to be loaded a second time into the rig to remove PCBs inaccessible in the original direction. An overview of this section of the disassembly process for both structure types is shown in Fig. 6.22.

For Type-I, the PCB cover is generally removed first, followed by the underneath PCBs. Considering the assembly direction, the PCBs are located above the carrier; the connections are easily seen, accessible, and can be disestablished from above (see Fig. 6.16a). The disassembly process is carried out until reaching the LCD module. On the other hand, for Type-II, the PCBs are generally connected to the carrier, which also covers the PCBs (see Fig. 6.16b). In this configuration the PCBs are inaccessible from above. After performing the operations that cut

```
START DISASSEMBLY
detectModel → unknown
detectBackcover → backCoverLocation = box_1
Treat Back cover (1)
  flagStateChangeROI(rect_1).
    op(c1,1): cutContour(rect_1', z_1)    SMF→ cutContour(rect_1', z_2) SMF→ cutContour(rect_1', z_3 ) SMF→
    op(c1,2): cutCorner(rect_1', z_1)     SMF→ cutCorner(rect_1', z_2) SMF→ cutCorner (rect_1', z_3)  SMF→
    op(c1,3): cutContour(rect_1'', z_1)   SMF→ cutContour(rect_1'', z_2) SMF→ cutContour(rect_1'', z_3) SMF→
    custom = [H_1]  →

          (Classification of 2 possible cases – The process can goes either Type-I or Type-II)
Type-I
(detectPcbCover → pcbCoverLocation = box_2) and (detectCarrier → pcbCoverLocation = 'no') → Type-I
Treat PCB cover(1)
  flagStateChangeROI(rect_2)
    op(c2,1): cutContour(rect_2',z1)      SMF→ cutContour(rect_2',z_2) SMF→ cutContour(rect_2', z_3)  SMF→
               cutLine(rect_2''^U, z_1)   SMF→ cutLine(rect_2''^U, z_2)  SMF→
          cutLine(rect_2''^D, z_1)        SMF→ cutLine(rect_2''^D, z_2)  SMF→
               cutLine(rect_2''^L, z_1)   SMF→ cutLine(rect_2''^L, z_2)  SMF→
               cutLine(rect_2''^R, z_1)   SMF→ cutLine(rect_2''^R, z_2)  SMF→
    custom = [H_2] →

detectPCBs → [pcbLocation = [box_3, box_4]
Treat PCB(1)
  flagStateChangeROI(rect_3)
    detectScrews → [screwLocation [loc_1, loc_2, ..., loc_n]] →
      op(cx,0): cutScrew(loc_1) → cutScrew(loc_2) → .... →cutScrew(loc_n)   SMF→
      op(c3,1): cutContour(rect_3', z_1)   SMF→ cutContour(rect_3', z_2)    SMF→ cutContour[rect_3', z_3)]  SMF→
      op(c3,2): cutCorner(rect_3', z_1)    SMF→ cutCorner(rect_3', z_2)     SMF→ cutCorner [rect_3', z_3)]  SMF→
      op(c3,3): cutContour(rect_3'', z_1)  SMF→ cutContour(rect_3'', z_2)   SMF→ cutContour[rect_3'', z_3)] SMF→
      custom = [H_3] →
Treat PCB(2)
  flagStateChangeROI(rect_4)
    detectScrews → [screwLocation = [loc_1, loc_2, ..., loc_n]] →
      op(cx,0): cutScrew(loc_1) → cutScrew(loc_2) → .... →cutScrew(loc_n)   SMF→
      op(c3,1): cutContour(rect_4', z_1)   SMF→ cutContour(rect_4', z_2)    SMF→ cutContour[rect_4', z_3)]  SMF→
      op(c3,2): cutCorner(rect_4', z_1)    SMF→ cutCorner(rect_4', z_2)     SMF→ cutCorner[rect_4', z_3)]   SMF→
      op(c3,3): cutContour(rect_4'', z_1)  SMF→ cutContour(rect_4'', z_2)   SMF→ cutContour[rect_4'', z_3)] SMF→
      custom = [H_4'] →

Type-II
(detectPcbCover → pcbCoverLocation = box_2) and (detectCarrier → pcbCoverLocation = 'box_5') → Type-II
Treat PCB cover(1)
  flagStateChangeROI(rect_2)
    op(c2,2): cutContour(rect_2',z1)      SMF→ cutContour(rect_2',z_2) SMF→ cutContour(rect_2', z_3)  SMF→
    custom = [H_5X] →

detectCarrier → carrierLocation = box_5
Treat Carrier(1)
  flagStateChangeROI(rect_5)
      op(c4,1): cutContour(rect_5', z_1)   SMF→ cutContour(rect_5', z_2)    SMF→ cutContour(rect_5', z_3 ) SMF→
      custom = [H_5] →

Treat LCD module(1)
      op(c5,1): cutContour(rect_1''', z_1) → cutContour(rect_1''', z_2) → cutContour(rect_1''', z_3) →

Post process
      custom = [H_6]
END DISASSEMBLY
```

Fig. 6.22 Disassembly process overview of **a** Type-I and **b** Type-II

around this covering area (*op*(*c2,1*)), this part of the carrier is separated from the remaining product with the PCBs still attached. In order to further separate the PCB, this part must be re-loaded into the rig in the opposite direction, in a second disassembly run after completion of the original process.

Carrying out this extra process is inefficient due to time consumption, however usually unavoidable for an unknown model. In a subsequent appearance of this model, the CRA will recognise this structure and try to improve the process. The strategy to achieve disassembly in a single run is to cut the screws from the back of the carrier. However, it is difficult for the VSM to detect the screws from this side due to their small size. A one-off human demonstration is required for the localisation of these screws. The CRA learns the demonstrated locations and executes this step automatically in subsequent processes. As a result, the PCBs can be also detached from the carrier in Type-II screens without reloading, with the consequence that the screws remain attached to the PCBs.

Unknown Model

In this example, the disassembly process at the state of removing PCBs is emphasised (see Fig. 6.22 for at the highlighted section). At the beginning of the disassembly process, the CRA requests the detection of the product model via comparison with the existing database. The result shows that the model is unknown; therefore, this disassembly is conducted as a trial-and-error process. The operations for the detected main components are executed according to each disassembly state. In the state that PCBs are expected, the CRA sends the sensing actions *detectPCB*. In reply, the VSM sends a fluent containing a list of PCBs: $pcbLocation = [box(x_{11}, y_{11}, x_{21}, y_{21}, z_{11}, z_{21}), box(x_{12}, y_{12}, x_{22}, y_{22}, z_{12}, z_{22})]$. The disassembly operation for the first PCB is started. The knowledge learned by the CRA, with regard to the operation in component-level, is represented by the highlighted fluents in Fig. 6.22. It can be seen that only the critical locations are collected from the general operation plans; meanwhile, all demonstrated action sequences from assistance are recorded.

Upon first entering the state, the original condition of the PCB is noted for benchmarking. This focuses on a rectangular ROI belonging to the PCB represented by a fluent $rect(x_{11},y_{11},x_{21},y_{21},z_{11})$.[8] The action used for observing the original condition is $flagStateChangeROI(x_{11},y_{11},x_{21},y_{21},z_{11})$.

Subsequently, semi-destructive disassembly is attempted via the disestablishment of screw connections. The CRA requests the sensing action *detectScrews*. Subsequently, the VSM returns a fluent containing the list of detected screws $screwLocation[(x_{11},x_{21},z_{11}), \ldots (x_{n1},x_{n1},z_{n1})]$. The remove screw operation *op(cs,0)* repeatedly performs the screw cutting operation at each location until the end of

[8] $rect(x_{11},y_{11},x_{21},y_{21},z_{11})$ is denoted by rect_1. The offset contour is denoted by rect'_1. The operation plan adjusts the cutting level to a particular depth z.

the list. The execution monitoring procedure is performed after each operation.[9] If the state changes, the system proceeds to disassemble the next PCB. However, an example treatment for the first PCB demonstrates the case that the component always fails to be removed by any general plan.

The general operation plan for PCB *op(c3,1)* uses the rectangle contour with offset as a cutting path and cutting method *m*. The cutting operation starts from the top surface at z_1 and incrementally cuts deeper in each operation cycle until z_3 where the boundary condition is reached. This operation is repeated for *op*(*c3*,2) and *op*(*c3*,3) with the respective offsets and cutting methods. If, after execution of all general operation plans, the PCB still fails to be removed, human demonstration provides a sequence of actions represented as *custom*. The complete information of demonstrated sequences is recorded, including the path *x, y, z* and a specific cutting method *m*. In this case, the demonstrated actions constitute cutting 2 lines, after which this component is detached.

Afterwards, the system proceeds to treat the second PCB. The example shows the successful removal of this PCB after the second trial of the general plan *op*(*c3*,1). The disassembly proceeds to the next state for treating the next main component. The system goes through the subsequent states until reaching the goal component, the LCD module. The relevant knowledge about the product and process has been learned and is now ready to implement when the same model is seen again.

Known Model

This section focuses on the knowledge in the KB that has been learned from previous disassembly process. In Fig. 6.23, the learning and revision in KB and the operation at each state are shown. This example shows the entire disassembly process whereby 6 main components are separated. Initially, during the first time disassembling this model, the CRA obtains the knowledge in both the product-level (see Fig. 6.23a: I) and component-level (see Fig. 6.23a: II–VII). This process is denoted as Revision-1.

After this model has been disassembled multiple times, significant changes have been made to the KB (Revision-5, Fig. 6.23b). Modification of the KB in the revision process occurs in three general categories: (a) *retraction of a plan*, (b) *learning of new component locations*, and (c) *learning of primitive cuts*.

Retraction occurs via removing the redundant general operation plans as described in Sect. 5.5.4, in effect making the process more efficient by reducing the time consumed. *New component locations* are demonstrated by the user to resolve errors in detection caused by the VSM. This process improves the accuracy of the cutting operation and the assessment of state change based on this location. *Additional primitive actions* can be demonstrated even when disassembly has previously been achieved in a given model. Extra cutting operations can resolve problems

[9] Regarding notation, the operation sequence is connected with the transition symbol $\xrightarrow{\text{<code>}}$. The code denotes the detail of the action executed in the transition state where $F = \mathit{flipTable}$, $S = \mathit{checkStateChange}$, and $M = \mathit{checkCuttingMethod}$.

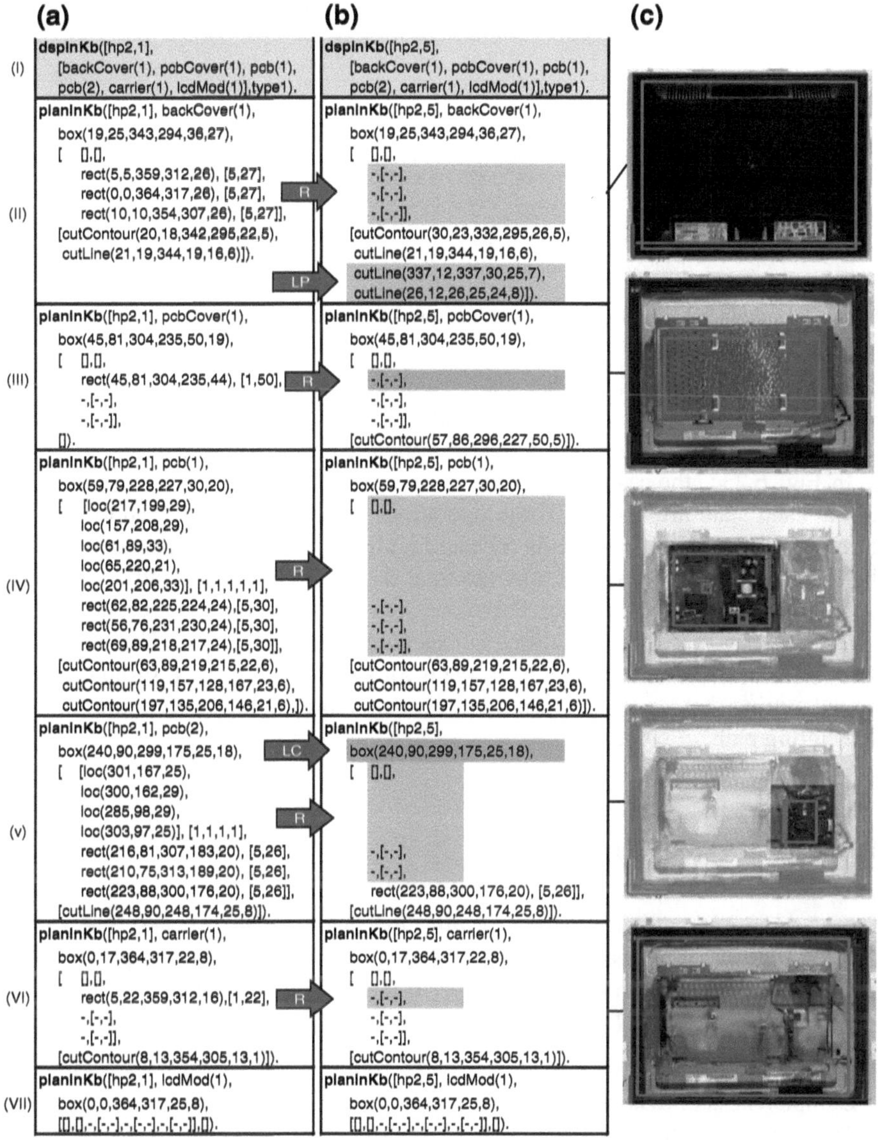

NOTE R = Retract the operations; LC = Learn new component location; LP = Learn primitive cuts.
LCD module was able to be detached without executing additional cut; therefore, no photos is shown in (c-VIII).

Fig. 6.23 Learning and revision

resulting from variation within the same model (e.g. modifications from the product usage phase), as well as uncertainties in the operation, (e.g. tool wear, melted plastic, inaccurate visual sensing of the cutting depth). Therefore, the process becomes more robust, should the same problems re-occur in the future.

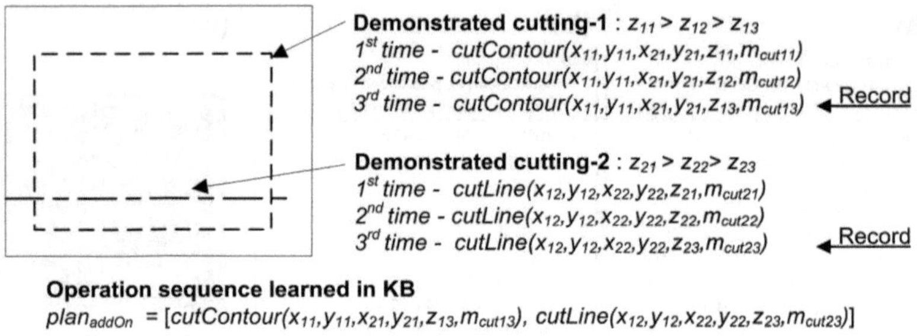

Fig. 6.24 Refining process for learning cutting operation for add-on plan

However, the operating time also slightly increases due to these additional actions. Through such revisions of the KB, the system is capable of improving its performance, hence reducing process time and the need for assistance.

In Fig. 6.23c, the *xy*-cutting paths of the human-demonstrated cutting operation at each state are shown. These operations are recorded in the added-on plan in the KB. The *xy*-cutting path is repeated multiple times at different depths since the human-operator does not know the final depth a priori. Trial-and-error is conducted, as directed by the user. Therefore, key primitive actions are only learned after the entire sequence of actions is executed (see Sect. 5.5.3). The learning of two demonstrated cuts (see Fig. 6.23c: I) is illustrated in Fig. 6.24. From 6 demonstrated cutting actions, only 2 primitive actions, representing one contour and one line cutting path, are recorded.

6.8.2 Key Performance Indices

This system was validated by performing the disassembly process. The experiments are designed to assess the performance of the system in two perspectives: (a) flexibility and robustness and (b) performance improvement, with respect to learning and revision. The assessments are done based on 3 key performance indices (KPI):

- Completeness of disassembly
- Time consumption
- Need of assistance

Completeness of Disassembly

Completeness of disassembly is considered in two aspects: (a) process completeness and (b) material separation efficiency. First, the *process completeness* describes the capability of the system to separate all desired components. The disassembly outcome is indicated as complete if every main component has been detached from one another. It also relates to the types of product structure where the number of the process needed is different.

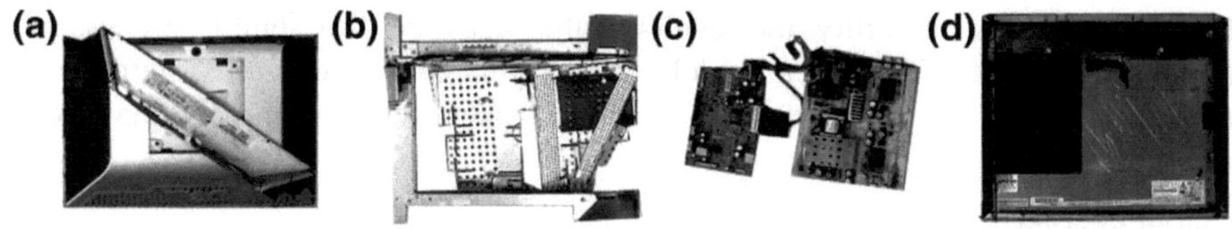

Fig. 6.25 Detached components classified by type of material. **a** Plastic. **b** Steel. **c** PCBs. **d** LCD module

The *material separation efficiency* describes the purity of the resultant separation of the various material categories. This metric is designed for (semi-)destructive disassembly where the output is in the form of lumps and scraps instead of complete parts. This is measured by *weight* in comparison with the *upstream* condition. Measurement is carried out after the disassembly process is completed.[10] The weight comparison is observed according to four groups of material: (a) plastics, consisting of the back cover and front cover, (b) steels, consisting of the PCB cover and carrier, (c) PCBs, and (d) a compound component, the LCD module. An example of the detached parts is shown in Fig. 6.25.

Time Consumption

Time consumption is commonly used for assessing system efficiency due to its direct relationship with operation cost. Time consumption is measured from the beginning to the end of each process, ignoring the lead time for the manual loading and unloading of the samples.

Need for Assistance

The amount of assistance required in achieving a task is an indirect measurement of the level of autonomy, which is more difficult to measure. Because each demonstration is a discrete application of just one motion primitive, the number of demonstrations is an objective measure of the amount of assistance required. The number of demonstration instances is counted from the beginning to end of the disassembly of each LCD screen.

6.8.3 Performance Testing—Flexibility of the System

The flexibility and robustness of the system describe its ability to handle the expected uncertainties and variations presented by previously unseen models. The behaviour of the CRA in handling unknown models is particularly relevant.

[10] The KPIs are measured at the end of the first run for *Type-I* and at the end of second run for *Type-II* structures. A constant penalty of 5 min is given when the PCBs-carrier connected part requires reloading in Type-II screens.

In order to assess flexibility, the results of the first-time disassembly process of a broad selection of LCD screen models[11] were collated. The overall system performance is affected by the performance of each operating module, summarised as follows:

Vision System
Performance in recognition and localisation is slightly reduced from the ideal case (see Table 6.3) due to the damage incurred on the main components from the (semi-)destructive disassembly method. The cutting operations occasionally destroy the common features necessary for recognising the components. As a result, recognition accuracy was generally reduced by 5–10 % in comparison with the ideal case. The localisation accuracy also changed within ±1 mm from the ideal case. The accuracy of the state change detection remained unchanged.

General Disassembly Plans
The general disassembly plans were executed autonomously to assess their robustness in handling unknown models. In theory, the parameters are developed based on the assumption that the vision system can provide accurate detected locations. Therefore, due to the overall position error of maximum 6.5 mm for *xy*-cutting path[12] and approximately 3 mm for the depth, complete component removal was sometimes difficult to achieve automatically. However, the required level of accuracy of vision system is hard to be achieved, so that additional operations, e.g. assistance, will also need to compensate this error.

Failures are caused by (a) inaccurate *xy*-cutting path, (b) insufficient depth of cut, and (c) non-detectable connections. Regarding the depth of cut, it is in some cases intentional to predefine a narrow range of cutting depths in order to decrease the chance of erroneously cutting through multiple components. For example, if the back cover is cut too deep, the border of the carrier may be also be cut at this state, resulting in the simultaneous detachment of the carrier, PCB cover, and PCBs, which are still connected to one another and need to be further disassembled in a second run. This decision results in a higher failure rate due to insufficient depth of cut.

It is obvious that the success rate of the PCB cover is quite high in comparison with other components. This can be attributed to the CRA's execution of the plan with multiple cutting offsets. This variation of cutting path compensates for the localisation error from the vision system. The high success rate is necessary, as the outcome is used to determine the product structure. The trial process

[11] The samples consist of 24 different selected models from 12 manufacturers. The size varied from 15″–19″, and the year of manufacture ranged over 1999–2011. With respect to the structure type, 15 monitors of Type-I and 9 monitors of Type-II were used. The Type-I sample was further classified into 8 of Type-Ia and 7 of Type-Ib.

[12] The *location accuracy* refers to the root mean square (RMS) error measured at the border. The error occurs on each side of the bounding box will be accumulated.

Table 6.5 Success rate of the plans for removing main components

Main component	Rate of success to remove the main component (%)				
	General plans conducted autonomously				Human assistance
	Plan-0	Plan-1	Plan-2	Plan-3	
Back cover	n/a	0	0	12.5	100.00
PCB cover	n/a	46.67	33.33	n/a	100.00
PCB	0	0	0	17.65	100.00
Carrier	n/a	16.67	n/a	n/a	100.00
LCD module	n/a				

costs extra operating time, however allows the system to learn the successful cutting path and implement a more efficient process in subsequent appearances of the model.

Overall, successful removal rates for the general plans were quite low but all components could be removed after assistance (see Table 6.5). Unsuccessful operation plans often indirectly contributed to the removal process by destroying the significant connections. The subsequent assistance allowed completion of the process.

Completeness of Disassembly

The system successfully identifies Type-I and Type-II via the proposed execution monitoring strategy (see Sect. 6.5.1). However, some misclassification occurs due to failure to remove the PCB cover part of the carrier. Approximately 70 % of all samples could be completed in the first run, with the remaining 30 % completed after the second run. The CRA directly learned the process for the first group. For the second group, the screw cutting strategy is taught in the subsequent revision and learned after that.

Regarding material output, around 97.36 % of the detached parts turned out as lump, with the remaining 2.64 % as scrap material. Considering the entire product, 98 % of the material was separated and collectible while other 2 % turned out to be small scraps and dust according to the cutting process. For each category, the efficiency of detaching plastic and the compound material (LCD module) were more than 93 % and the efficiency of PCB and steel were around 85 %. One of the critical issues in disassembly of LCD screens is the risk of damaging the Cold Cathode Fluorescent Lamps (CCFL) lying within the LCD module. From this experiment, minor damage occurred to the LCD module in most cases due to the cutting operations for the carrier. However, due to the predefined constraints, there was no single case where the CCFL were damaged. Therefore, it can be concluded that no toxic substances were leaked from the CCFL.

Time Consumption

The duration of the process depends on the complexity and the size of the screen. On average, around 97 % of the total time consumed was due to physical

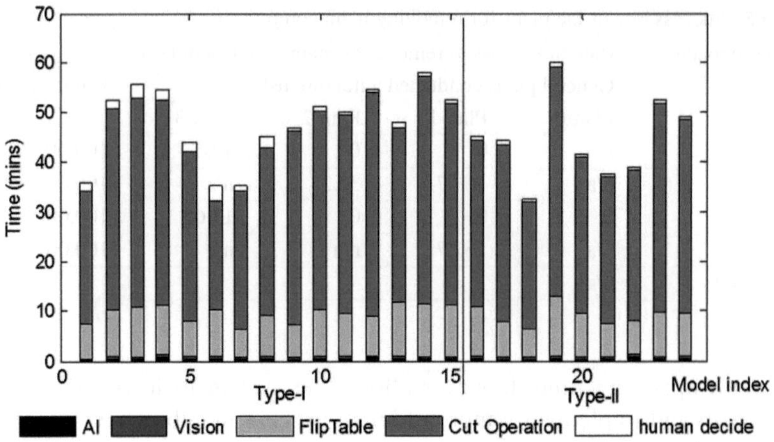

Fig. 6.26 Time consumption of the disassembly process by each operation

operations,[13] 1.5 % due to visual sensing and artificial intelligence (AI),[14] and 1.5 % due to data transfer activity. The autonomous trial-and-error process performed 67 % of the entire disassembly process while the remaining 33 % was directed by human demonstration.

The time required to complete an entire disassembly was on average 48 min, varying between 32 and 60 min according to the complexity of the samples. This process time is very long in comparison to traditional manual disassembly. However, the process for each individual model is expected to be further optimised via the learning and revision strategy. Firstly, the redundant operations for trial-and-error and learning will be reduced: the flipping table routine, reloading for second run and assistance. The time spent by each type of operation is clearly shown in Fig. 6.26 where most of the flipping table routine and cutting operations were redundant. At least an average of 11–16 min will be regained by removing these redundancies. Successful cutting methods will also be known, further reducing the time required testing cutting methods and recovering from collisions. Secondly, the redundant cutting operations will be retracted. Lastly, as the cutting depth is known, larger increments can be taken in order to make the process

[13] The operation routines in a operating cycle:

- Flipping operation (8.45 s) and checking state change (2.45 s);
- Updating the grinder's size: movement and visual checking (4.19 s); and,
- Cutting operations (average 33.75 s). It varied from 5.09 to 186.91 s (95 % of the data were within 0–100 s) depending on the size of the cutting path and variation of the cutting method trial.

[14] The activity of the AI includes the decision-making process of the human operator during demonstration steps: approximately 5 s per demonstrated action.

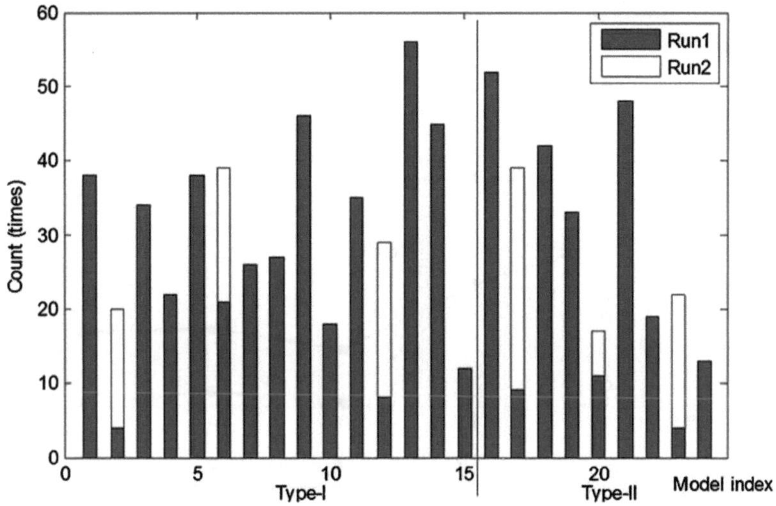

Fig. 6.27 Assistance count in the disassembly process

quicker. The time regained is dependent on the characteristics of the operations required. From results in the following section, the time consumption is finally reduced to around 24 min.

Need for Assistance

99.9 % of the assistance provided consisted of the demonstration of a primitive cutting path. Less than 0.1 % was for correcting an error in state or component detection. The overall amount of assistance for each model is shown in Fig. 6.27 where *Type-I* and *Type-II* are shown separately. The counts for *Type-I* and *Type-II* were similar, averaging 32 instances and ranging from 12 to 56. The majority of counts in *Type-II* occurred due to the demonstration of the strategy for cutting screws and/or during second runs.

Summary

In conclusion, the general operation plans were capable of accomplishing most of the required work in disassembling unknown models, however were unable to complete the process without assistance. Assistance resolved all remaining uncertainties, compensating for the limitations in the perception of the system. Since component separation was achieved in every case, it can be concluded that the disassembly is achievable via the physical configuration and vertical cutting direction. However, the time required needs to be improved in order to compete with the traditional manual disassembly. Even though the operation was not fully automatic in the first-time disassembly of unknown models, the process was learned. The system is expected to increase in its degree of automation with more experience disassembling each given model.

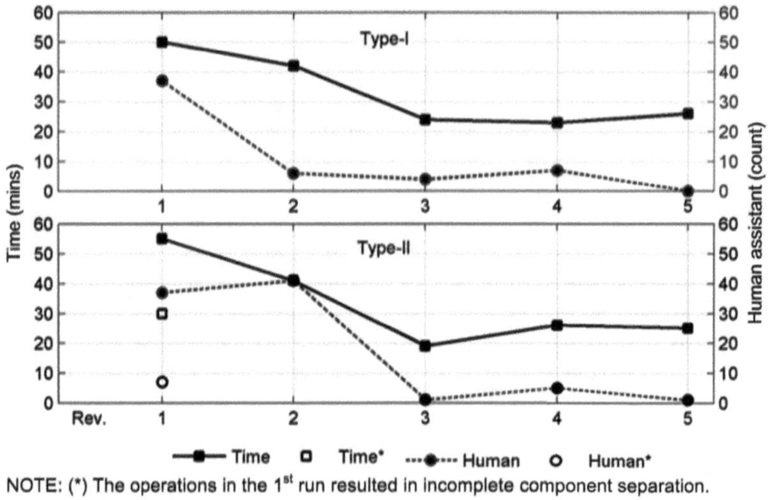

Fig. 6.28 Disassembly performance with respect to multiple revisions

6.8.4 Performance Testing—Learning and Revision

This experiment aims to assess the capability of the system both in learning and in revising the knowledge extracted from previously-executed disassembly processes. These behaviours are expected to improve the performance of the disassembly process. The experiment was executed on two models that embody the typical characteristics of the two structure types. Each model was disassembled five times, each time building upon the knowledge learned from previous disassembly processes, in order to see the trend of performance improvement due to learning and revision.

Results showed a dramatic improvement in the performance of the system during the first few revisions, with regard to time consumption and the amount of assistance required. System performance then levels out, remaining roughly constant with small fluctuations due to the variability in the vision system and the disassembly operation processes. These trends can be seen in Fig. 6.28. Detailed observations regarding each model are explained in the following. Significant differences are observed between the two structure types.

Type-I Structure

The entire disassembly process is expected to be completed in a single run. This condition was satisfied from the first-time disassembly of this model. In the first disassembly (*Rev-1*), the overall time consumption was 47.9 min with 37 instances of assistance. Around 70 % of the assistance was given for deepening cuts that were insufficient to remove the components. The rest were used to locate non-detectable connections. The revisions were done later on until *Rev-5*. In comparison to the first disassembly, the time consumption decreased to about

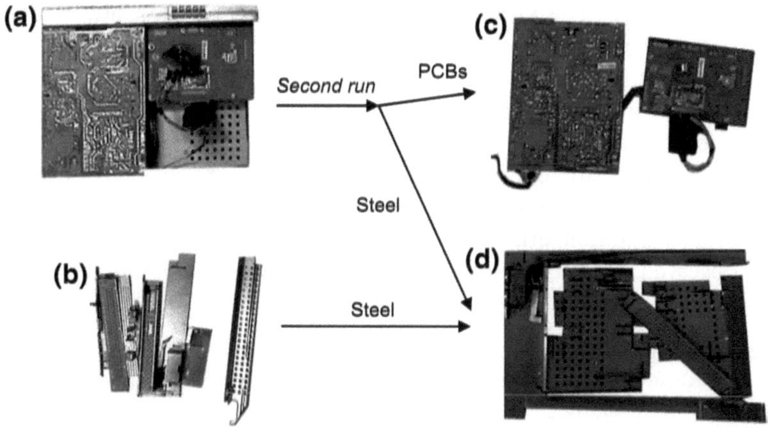

Fig. 6.29 Incompletely detached carrier and PCBs and the second run. **a** Incompletely detached PCB and carrier. **b** Cut-off steel part from carrier. **c** Completely detached PCBs. **d** Completely detached carrier

89 % in *Rev-2* and 52 % in *Rev-3*. Small fluctuations of within 5 % occurred between *Rev-3* and *Rev-5*. A similar trend occurred regarding assistance. The assistance count dropped dramatically to 16 % in *Rev-2*, maintained this level within 3 % fluctuation, and then finally dropped to 0 % at the *Rev-5*.

In summary, the performance increased as the process was revised. The final disassembly time was 25.7 min and no further assistance was needed in the final test. There was no significant difference in the material separation efficiency amongst the revisions, which remained at approximately 98 %.

Type-II Structure

Due to the configuration of the PCBs under the carrier, a second run was initially needed to complete disassembly (*Rev-1*). However, this problem was resolved in the second-time disassembly (*Rev-2*), where only a single run was needed, since the strategy for cutting screws from outside the PCB cover had been taught. The incompletely detached carrier and PCBs, and the material outcome after performing the strategy taught in the second run are shown in Fig. 6.29.

In the first revision, the total time consumption (sum of the first run, second run, and reload penalty) was 28.1 + 22.0 + 5.0 = 55.1 min. The total amount of assistance in the first and the second runs was 8 + 30 = 38 instances. As with the *Type-I* case, most assistance in *Rev-1* was given for deepening existing cuts. In *Rev-2*, the CRA realised that multiple runs were previously required—a suboptimal solution. Consequently, it asked for the user's assistance for employing the screw-cutting strategy. As a result, the time consumption decreased to 74.5 %. However, more counts of assistance were required, increasing to 107.9 % of the first run. The problem of cutting depth was almost completely resolved during *Rev-I*; more than 90 % of assistance in *Rev-2* were given to direct the screw-cutting strategy since the VSM is unable to detect the screws in this circumstance. The knowledge obtained in this revision now satisfies the more efficient single-run requirement. In *Rev-3*, both values decreased dramatically, whereby time consumption decreased to 34.5 % and assistance to 2.6 % according to the first revision. Both values marginally increased by about 10 % in *Rev-4* due to process uncertainties (see following section) and maintained about this level until the end of *Rev-5*. At the final revision, the time consumption was 25.1 min (45.5 % of the first revision) and no assistance was required. The material separation efficiency was around 90 %, with no significant difference observed between the revisions. The disassembly outcome is shown Fig. 6.25.

Uncertainties in Process

The results showed a fluctuation in performance in some revisions, e.g. *Type-I Rev4-5* and *Type-II Rev3-4*. These fluctuations were caused by the uncertainties in vision system and disassembly operation.

Firstly, significant uncertainty in the vision system is caused by the inaccuracy of *measureZF*, used to locate the top surface for cutting. The precision of *measureZF* suggests that the measured top cutting level lies within approximately ±3 mm from the actual top surface. As illustrated in Fig. 6.30, this inaccuracy may cause extra deepening iterations to be required. The CRA cuts the object from the start point to a destination depth z_{dst} learned from the previous operation. When the sensed top surface is too high, an extra cut is necessary (see Fig. 6.30b). This extra cut results in extra operating time. This variation is limited by the precision of the sensor.

Secondly, the disassembly operation involves an imperfectly-repeatable cut. An identical action, successful in detaching a component in a previous process, may fail in subsequent processes. The problems are caused by two factors of uncertainty: (a) non-uniform wear rate of the abrasive cutter, resulting in minor differences in cutting location, and (b) uncertain physical responses of materials to different process parameters. For instance, plastic may be melted when using a higher feed speed or larger cutting depth, causing components to become stuck together. Extra assistance is required to resolve these minor uncertainties, with extra time consumption as the consequence.

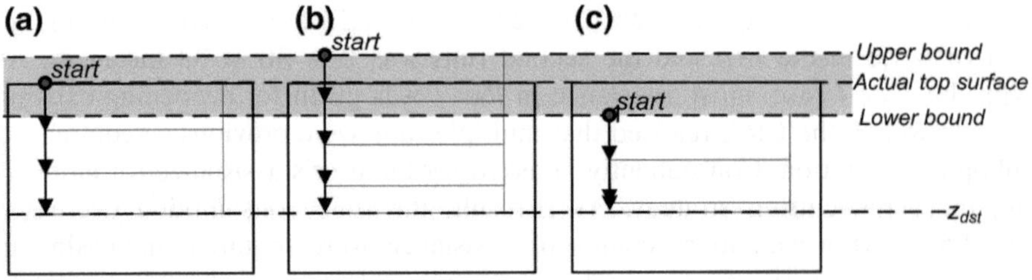

Fig. 6.30 Uncertainties due to the variation of the starting level for cutting. **a** Ideal case. **b** Too high measured z. **c** Too low measured z

Summary
In conclusion, due to the implementation of learning and revision, the performance of the system increases to a certain level as more disassembly processes are conducted on each model. This performance improvement is particularly seen with regard to the Type-II structure, which can be completed with a single run in the second and subsequent revisions. The time consumption of the disassembly process decreased dramatically in the first few revisions, to a final disassembly time of about 25 min, approximately 50 % of the time for the first run. After the revised KB becomes stable after the first few revisions, the process is expected to be performed autonomously without any or with only minimal assistance. Minor uncertainties in the process cause the performance to fluctuate within bounds. This is expected to be further suppressed by improving the accuracy of each individual operating module.

6.8.5 Conclusion and Future Improvement

These experiments validated the performance of the system, proving it to be flexible and robust to the variation and uncertainties found in the case-study EOL products. The autonomous process together with some assistance was able to complete the disassembly of every given sample. In addition, it has been proved that the performance of the system can be improved via the learning and revision strategy. After a few revisions, the system is able to complete the process within a shorter time and fully autonomously in good conditions. These cognitive abilities are important for the development of an automated disassembly system where flexibility, robustness, and cost minimisation are crucial. This system has successfully proved the principle that cognitive robotics is able to overcome the uncertainties and variations in the disassembly process.

However, in comparison to traditional manual disassembly, the process time consumption is still too high, rendering the system economically infeasible. The time consumption needs to be further reduced to 6.2 min/screen through the prospective improvements which are as follows. Firstly, the disassembly time can be reduced to 9–11 min if the cutting tool can approach the destination depth in one operation cycle instead of incrementally deepening the cuts. More powerful and reliable hardware is required for this purpose. Secondly, all physical movements can further be optimised: e.g. feed speed, robot movement, flipping table, reducing the time required for each primitive action.

Moreover, the dependence on the human operator should be minimised, since this is directly associated with operation cost. The general plans should be improved to become more capable of carrying out the process autonomously. Even though the proposed concept of common location of the connectors is able to identify an appropriate border offset, visual localisation must be improved in order to achieve a higher success rate.

6.9 Conclusions

In this chapter, an automated disassembly system with cognitive abilities is described in detail. This system is designed for disassembling LCD screens, selected as the case study product. To improve economic feasibility, (semi-)destructive selective disassembly is implemented. For flexibility and robustness, the disassembly strategy, the principle of cognitive robotics and assistance together address the uncertainties and variations in products and process.

Firstly, a significant number of LCD screens were examined. The typical variations and uncertainties in the products were found in the different models of LCD screen. It was found that the product can be classified into 2 main structure types, *Type-I* and *Type-II*, according to the configuration of the main components. Typically, LCD screens consist of 6 main components and 4 connective components. This information was used in designing each operating module in the system.

The tasks, designed to handle the types of uncertainties expected in the process, are distributed between three operating modules. The CRA is the high-level planner that controls the disassembly process according to the desired cognitive behaviours. This addresses the variations in the product structure and associated disassembly plan. Using this methodology, the disassembly process can also be learned and improved later on. The second module, the VSM, recognises and locates the relevant main and connective components. This module takes into account the variations in physical appearance of components in different models. The third module, the DOM, performs the physical operations on the samples. The general operation plans are designed to address uncertainties in the cutting operations where the visual information and the related knowledge are not precise. Moreover, the human operator uses this module to perform the demonstrated actions, allowing the system to complete disassembly when the automatic process fails.

Finally, the operating modules not only perform their own specific tasks but also help in addressing the unresolved problems of other modules. For example, in case of the connective components occasionally misdetected by VSM, the general plans in DOM are performed multiple times according to the CRA's decision to increase the success rate of component removal. In conclusion, as a whole, the system was able to experimentally demonstrate its ability to automate the disassembly of a significant number of different models of LCD screens.

References

1. Kernbaum S, Franke D, Seliger G (2009) Flat screen monitor disassembly and testing for remanufacturing. Int J Sustain Manuf 1(3):347–360
2. Ryan A, O'Donoghue L, Lewis H (2011) Characterising components of liquid crystal displays to facilitate disassembly. J Clean Prod 19(9–10):1066–1071

3. Parliament (2003) Directive 2002/96/EC of the European Parliament and of the council on waste electrical and electronic equipment (WEEE) of 27 Jan 2003
4. Franke C, Kernbaum S, Seliger G (2006) Remanufacturing of flat screen monitors. In: Brissaud D, Tichkiewitch S, Zwolinski P (ed) Innovation in life cycle engineering and sustainable development, pp 139–152
5. Kim HJ, Kernbaum S, Seliger G (2009) Emulation-based control of a disassembly system for LCD monitors. Int J Adv Manuf Technol 40(3–4):383–392
6. Uhlmann E, Friedrich T, Seliger G, Harms R (2005) Realization of an adaptive modular control for a disassembly system. In: The 6th IEEE international symposium on assembly and task planning: from nano to macro assembly and manufacturing
7. Reese G (2000) Distributed application architecture. In: Database programming with JDBC and Java, 2nd edn. O'Reilly & Associates
8. MSDN M (2011) Windows sockets 2. http://msdn.microsoft.com/en-us/library/ms740673(v=vs.85).aspx
9. Viggiano JAS (2004) Comparison of the accuracy of different white balancing options as quantified by their color constancy. In: SPIE—The International Society for Optical Engineering, pp 323–333
10. Chang F, Chen C-J, Lu C-J (2004) A linear-time component-labeling algorithm using contour tracing technique. Comput Vis Image Underst 93(2):206–220
11. Viola P, Jones M (2001) Rapid object detection using a boosted cascade of simple features. In: IEEE Computer Society conference on computer vision and pattern recognition, pp 1511–1518
12. Bay H, Ess A, Tuytelaars T, Van Gool L (2008) Speeded-up robust features (SURF). Comput Vis Image Underst 110(2):346–359
13. Vongbunyong S, Kara S, Pagnucco M (2015) General plans for removing main components in cognitive robotic disassembly automation. In: The 6th international conference on automation, robotics and applications (ICARA 2015), Queenstown, New Zealand, 17–19 Feb 2015

Chapter 7
Conclusions

Abstract This chapter presents the final conclusions regarding disassembly automation in general and the cognitive robotic disassembly system developed in this research. The technical aspects of this research, the results and the lessons learned are summarised. In light of this, our conclusions and direction for future work are also presented.

7.1 Conclusion in Technical Perspectives

According to the principles of disassembly automation presented in Chap. 3, the system consists of three primary elements: the disassembly operation unit, vision system, and intelligent agent. In this research, these take the form of the disassembly operation modules (DOM), vision system module (VSM), and cognitive robotic module (CRM), respectively. The steps taken, and lessons learned in the implementation of these modules are summarised in the following sections. Finally, the method in which the system addresses the problem of various uncertainties, which has thus far hindered the implementation of disassembly automation, is addressed.

7.1.1 Disassembly Operation Module

The disassembly operation module deals with physical operations conducted in the disassembly process. The development involves three primary parts, including product analysis, hardware design, and operation plans.

Product analysis
In general, product structure information of specific models is required to select disassembly operations and an optimal disassembly sequence plan. Unfortunately, it is often the case that EOL products are returned where the specific information,

S. Vongbunyong and W.H. Chen, *Disassembly Automation*, Sustainable Production, Life Cycle Engineering and Management, DOI 10.1007/978-3-319-15183-0_7

e.g. CAD model and assembly sequence, is unavailable. Such information can be generated from the examination of select models; however the effort to research and implement for a specific model tends to be unjustifiable in industry practice, where the products returned are in great variety and the lot sizes are unpredictable.

In this research, the product family is analysed in order to identify the primary variations between models. The CRA is programmed to address these uncertainties and variations, eliminating the need for a priori model-specific information. A broad idea of the main product structure is given to guide the process; the concepts used are general enough to cover the variations between models but not too broad such to limit the size of the search space. In this case, the order of the main components in LCD screens is quite consistent; only two types of main structures are defined. The types and characteristics of the components—main and connective—also need to be identified in order to design the operation plans and required disassembly tools.

Mechanical system

The mechanical system as the disassembly operation unit is designed according to these requirements. In general, the current research trend focuses on two perspectives which are (a) the development of the entire disassembly system and (b) the development of specific disassembly tools for removing types of connectors as presented in Chap. 3. The primary components of the automatic disassembly workstation are robot arms equipped with disassembly tools, grippers, and the fixture system.

Automatic disestablishment of fasteners generally requires specifically-designed tools and accuracy of actions. Greater accuracy can be attained with force-torque control and/or closed-loop visual feedback. Detachment of main components is generally achieved using grippers. Complexity usually arises according to the variable geometry of the components and the requirements to create a firm grip. However, physical uncertainties are found in both cases resulting in difficulty in the removal process. If the failure of an operation can be identified, the system may be able to recover by appropriately executing an alternate control sequence.

In this research, the (semi-) destructive approach is employed for a higher success rate in the midst of uncertainties, requiring lower accuracy in force-torque and position control to achieve the goal of component separation. In addition, the complexity of the gripping and fixture system is eliminated by using the flipping table, which is able to remove detached components without any knowledge of their geometry. The proposed design simplifies the process and reduces the setup cost of the system.

Operation plans and process parameters

Where this information of a particular product is known in advance, the operation plans and parameters can be predefined. However, this is often infeasible as discussed earlier. The method of formulating and utilising operation plans and parameters is one of the primary contributions of this research. Only the broad scheme of operation plans and parameters are supplied, as choice points in the space that the CRA searches while carrying out the disassembly process in a trial-and-error manner. As a result, the specific information of the product does not need to be supplied a priori.

In this research, the general operation plans for each type of component is formulated from statistical information of possible locations of the corresponding connective components. Given that the partially damaged disassembly outcome is acceptable, instead of destroying the connective component (the semi-destructive approach), the most effective operations consist of cutting near the border of the main component (the destructive approach) to detach the majority of its material. A major benefit is to compensate the overhead vision system's inability to detect fasteners such as hidden snap-fits and screws located around the sides. This method is also able to identify the process parameters, e.g. the critical depth of cut for successful removal of a section, which is also unobservable from the outside.

7.1.2 Vision System Module

Vision is generally the primary sensing method in disassembly automation. Its main function is to detect—recognise and localise—the components in the disassembly process. Sensing capabilities are crucial for the disassembly process since certain information is only disclosed during the disassembly process. Therefore, the current condition of the process needs to be updated for appropriate decisions to be made by the planner. In this research, the vision system module is able to address the uncertainties of physical appearance, quantity, and location of the component. The vision system needs to be considered in two perspectives: hardware capability and algorithms.

Hardware capability
A number of techniques with different advantages and limitation are currently used for sensing the information in 2D, 2.5D, and 3D, as discussed in Chap. 4. The techniques must be selected according to the requirements of the disassembly process. In this research, the colour and the depth camera is used for generating a 2.5D image map. This selected option is more cost effective and consumes less computational resources in comparison to other techniques. Data loss occurred in the infrared-based depth sensor at reflective surfaces perpendicular to the infrared emitting direction. Additionally, accuracy is reduced at the edges of the object. These problems are partly eliminated by a filtering algorithm that disregards the irrelevant data. The inaccuracy at the edges is taken into account by setting appropriate cutting offsets in the operation plans.

Detection algorithms
There is no general solution that is effective for detecting every type of component. A detector needs to be developed targeting particular types of components, in most cases, by combining a number of machine vision and pattern recognition techniques. The problem of detection becomes more difficult due to the

unpredictable EOL condition, e.g. damage, partial occlusion and missing parts. The detector must also be able to address these issues.

In this research, the rule-based recognition of the components using the concept of *common features* addresses these variations in appearance. This recognition technique is one contribution of this research. Predefined rules are developed according to the common physical appearances of each type of component observed from the samples. The developed algorithms are able to accurately recognise the type of component and the location. This is flexible enough to deal with the variations between different product models of products, however shows varying performance according to the degree of damage to the component.

Another important function of the vision system in this research is to assess the success of the operations. This assessment is part of the execution monitoring which is done by checking for a change of disassembly state. This measurement is designed to support destructive disassembly, in which success is defined as the removal of a significant part of the component, rather than its entirety. The detector is developed based on similarity measurements based on both the colour and depth images. This has been shown to achieve 95 % accuracy in determining state change in this application.

7.1.3 Cognitive Robotics Module

The intelligent agent controls the automatic disassembly system to perform the disassembly process. To make the system robust and flexible, the agent needs to take existing knowledge and plans and adapt it according to the currently-sensed information. A number of examples can be seen from the existing research work associated with adaptive planning in Chap. 2 and the disassembly systems described in Chap. 3.

In this research, the cognitive robotic agent controls the system using four cognitive functions: (a) reasoning, (b) execution monitoring, (c) learning, and (d) revision. The uncertainties regarding product structure and process are addressed by this agent. The CRM is composed of the CRA and the KB. The CRA controls the behaviour of the system and the KB contains the model-specific knowledge that is obtained from the previous disassembly processes. A key benefit of using cognitive robotics in this research is the ability to make decisions according to actual execution outcomes. The system improves from previous experience by learning new instructions given by the human operator. These features give the system flexibility to deal with various models of products, by addressing the uncertainties at both the planning and operational levels.

Architecture and language platform

The cognitive robotic architecture is based on the closed perception-action loop which expresses the key features of the behaviour, i.e. perception, action, reasoning, learning, planning, behaviour control, and human interaction. The CRA is programmed with action programming language *IndiGolog*, which is based on Situation Calculus. The key benefit to this research is the online execution which supports sensing and exogenous actions which allow the CRA to effectively respond to the external dynamic world. The language of the *Golog* series also benefits the development process, as the behaviour of the system can be clearly described by actions, preconditions, and effects.

Reasoning and execution monitoring

The CRA schedules actions by reasoning about the current condition of the disassembly state, the disassembly domain, and the execution outcome. The existing model-specific knowledge is also taken into account for disassembling the known models. In addition, the CRA can decide to switch to user assistance when the autonomous operations have failed too many times.

For the *unknown model*, the key feature of reasoning is to select the operation plans and parameters according to the current main component. They are considered choice points to be pruned along with the two main structure definitions. The input is obtained by the component detectors and the execution outcome which is determined by the execution monitoring that examines the change of disassembly state. This input is used in the trial-and-error process in order to find the critical plan and parameters that lead to the state change. As a result, this can eliminate the need of the disassembly sequence plan (DSP) and disassembly process plan (DPP) supplied a priori. The CRA will also learn these generated DPP and DSP. This also addresses the uncertainties due to the variation in the quantity of the components, e.g. of PCBs. In the case of the *known model*, reasoning is used to execute the operation according to the knowledge previously learned. The sensing input in regard to the component type and location is less significant in this process since the information is already known.

The CRA performs the disassembly according to the order of the states defined by the given main structure. These predefined structures benefit the reliability of the process by reducing the effect of the misclassification of main components, which is more frequent when the components are damaged. Effects of misclassification include infinite execution loops and redundant physical damage, leading to increased time consumption and the learning of incorrect information. However, a major drawback of using given broad structures lies in the inability to handle a product whose structure differs significantly from the given definition. This may be the case in other product families where the structure is more complex. However, this has not been found to be a problem in the observed case-study products.

Learning and revision

In this research, learning occurs in two forms. First, in *learning by reasoning*, the CRA learns the parameters for the predefined general operation plans that have been executed prior to the successful component removal. The critical values of all executed operations need to be recorded even if the state change does not immediately occur, since some cuts may passively contribute to the detachment. Second, in *learning by demonstration*, the CRA learns from the assistance given to overcome unresolved problems. Assistance may be given to change original beliefs caused by inaccurate visual detection, e.g. regarding the existence of a main component or the occurrence of state change. In addition, assistance is given in the form of additional sequences of primitive cutting operations (custom plans), which may disestablish the remaining connections that are non-detectable or require deeper cuts.

A major benefit of learning is to reduce the need for assistance for disassembling known models. The time consumption is also marginally reduced by skipping redundant steps, e.g. flipping the table and visual sensing. A limitation of the described strategy is that knowledge cannot be adapted between different models. Therefore, specific information is generally supplied by the human user during the first disassembly of every unknown model. However, the experimental results prove that the CRA requires significantly less human intervention in subsequent appearances of the model, due to learning and executing taught steps autonomously.

In the *revision* process, the disassembly process of known models can be optimised by retracting the redundant general operation plans that have been learned previously. Redundant operations, which do not contribute to the removal of the main component, can be found by executing the operation plans in an alternative order and checking for the detachment of the component before all actions in the plan have been executed. A simplified form of this concept is implemented in this research, by reversing the order of operations in the plan.

Experiments show a significant improvement in process efficiency after a small number of appearances of the same model. The time consumption reduced by more than 50 % from the first appearance and the process was able to be carried out without assistance after the first few revisions. However, small fluctuations occurred due to inaccuracies in the visual localisation and physical operation.

In conclusion, the performance of the system significantly improves in every aspect after the first few tries to disassemble a particular model. This is due to the learning and revision strategy which is able to obtain the necessary specific model information during online operation. Subsequently, the process can be conducted largely autonomously and in a robust manner. This strategy has not previously been trialed in other research work.

7.1.4 Flexibility to Deal with Uncertainties

In this research, a system has been built to deal with the uncertainties that have prevented the implementation of disassembly automation for the treatment of EOL electrical/electronic waste. The uncertainties can typically be addressed autonomously by the integration of the operation modules. Principles that have been applied in this system include:

- identifying broad, abstract product structures that describe the entire range of known models;
- detecting components by type;

- supplying and pruning the search space of a range of possible operations, created in consideration of the entire product range; and
- learning from assistance, which is only provided in cases that are unresolvable by the autonomous system.

The primary uncertainties listed in Table 3.1 are discussed as follows.

First, the *uncertainties in EOL condition* that cannot be observed by the vision system are expected to be addressed by the CRM and DOM. As a result, the cutting location cannot be determined accurately. The general operation plans, which are part of DOM and used by CRM, cut the main component at the estimated location that expects to disestablish the connection. The force-torque feedback is used to acknowledge collisions and find alternative cutting method to achieve the operation.

Second, the *diversity of the supplied products* concerns the variation in component appearance, quantity and location between models of the same product. This is addressed by detecting components by type, as opposed to using a specific template. The broad structure category that each sample falls under is identified by the CRA according observations during the disassembly process. The success of addressing these uncertainties depends on the performance of the vision system.

Third, uncertainties in the required *process and operations* plan is typically addressed by the CRA, which uses reasoning and execution monitoring to obtain effective sequence plans, operation plans, and process parameters through trial-and-error. Errors due to sensor or actuator limitations are addressed in the same way as uncertainties in the EOL condition, compensated by the trial-and-error strategy or assistance. The success of addressing these uncertainties depends on the accuracy of the predefined structure and operations, and the constraints of the process parameters.

The ability of the system in dealing with these uncertainties has been validated experimentally on numerous models of LCD screens. The success rate of the automatic operation can be increased if less strict constraints are assigned e.g. deeper maximum cutting depth. This will lead to a greater potential for the trial-and-error approach to complete the task, at the expense of increased time consumption. The amount of uncertainty and the need for assistance reduces significantly after the model-specific knowledge has been learned.

The integration of these principles is a starting point in bridging a gap in existing research work, where previously only known models can be disassembled by an autonomous system.

7.2 Economic Feasibility

In practice, one of the major problems of disassembly in industrial scenario is the unpredictable quality and quantity of the products returned. The fully automatic disassembly system can be economically feasible if:

- The system supports a wide range of products;
- Enough uncertainties can be addressed autonomously;

- The process is reasonably fast;
- Human operators' direct exposure to hazards can be avoided; and,
- High value can be recovered from the disassembly outcome.

The system developed in this research addresses some of these issues. Economic feasibility is preliminarily evaluated in two perspectives, including the cost of the automation platform and the operating cost.

First, a low-cost platform that is flexible to deal with a wide range of models of LCD screens has been designed in this research. Using the destructive approach and specially designed tools, the system achieves a high success rate of disassembly. Regarding value recovery, the damaged condition of the disassembly outcome is suitable only for recycling. The system should aim for non-destructive or semi-destructive disassembly to acquire higher value returned.

Second, time consumption is a key concern which directly relates to the operating cost. In this, the current prototype system still needs further improvement. In comparison to a comparable manual process which takes 6.2 min/screen on average [1], the proposed system takes approximately 48 min for an unknown model sample, which reduces to around 24 min after a few revisions. Improvements in regard to physical operation and hardware are needed to overcome this limitation. It is a goal for the disassembly time to be reduced to less than 10 min.

7.3 Conclusions of the Research

7.3.1 Conclusions

In regard to the development of disassembly automation, the flexibility to deal with various models of products is crucial for the industrial application. This research shows the possibility of making disassembly automation economically feasible by using the concept of cognitive robotics together with the associated operating modules. In addition, learning and revision are key features that allow the system to improve the process performance from previous experience. Even though the human needs to be involved in the first stages upon receiving an unknown product, the system learns from this to become more autonomous.

7.3.2 Future Work

Regarding future work, the system should be further developed towards economic feasibility. A primary goal is to bring the physical system near the efficiency and flexibility of manual disassembly through the improvement of the hardware and the optimisation of operations. In addition, the following additions are foreseen for extending the system to support a wider range of products.

The basic behaviour of the cognitive robotic agent will become more complex, in order to address the uncertainties in a wider range of products. A limitation in the current learning and revision strategy is that newly-generated knowledge is specific to individual models. Therefore, a strategy that allows the robot to adapt the existing knowledge of one model to disassemble another model should be developed. This implementation can take the form of learning broad rules relating learned operations or parameters to observations, so that with experience, the system also increases in its ability to autonomously handle unknown models.

The mechanical system will be extended to include a variety of disassembly tools and grippers to facilitate the non-destructive disassembly of select components. This is desirable for increasing the economic gains from the disassembly outcome, since undamaged components can re-enter the product stream as spare parts or for remanufacturing. It is desirable for aspects of the system to be modular; system components such as the product fixture may need to be re-designed to hold products from a different family. Increased reliability is desirable for both the disassembly operation unit and the vision system.

Reference

1. Kernbaum S, Franke D, Seliger G (2009) Flat screen monitor disassembly and testing for remanufacturing. Int J Sustain Manuf 1(3):347–360

Appendix A
Actions and Fluents

See Tables A.1, A.2 and A.3.

Table A.1 Sensing actions and corresponding fluents

Sensing action	Fluent	Description
detectBackCover	backCoverLocation	Locate back cover
detectPcbCover	pcbCoverLocation	Locate PCB cover
detectPcb	pcbLocation	Locate PCBs location
detectCarrier	carrierLocation	Locate carrier
detectLcdModule	lcdModule	Check existence of LCD module
detectModel	Model	Match the model of the sample with the models in KB
checkStateChange	stateChange	Determine change of the state
measureZF	currentZF	Measure level-z
SenseHumanAssist	humanAssistOperation	Get assistance from human
checkCutting-Method	cuttingMethod	Get cutting method from the robot controller

Table A.2 Primitive actions and corresponding fluents

Category	Primitive action	Fluent	Description
Primitive cutting operation (1)	*cutPoint*	*loc(x,y,z) & m*	Cut point, e.g. screw
	cutLine	*line(x,y,x,y,z) & m*	Cut straight line with cutting method-m
	cutContour	*rect(x,y,x,y,z) & m*	Cut a contour with cutting method-m
	cutCorner	*rect(x,y,x,y,z) & m*	Cut corner with cutting method-m
Disassembly operation utility	*flipTable*	–	Activate flipping table
	moveHome	–	Move to robot's home
	flagStateChange	*stateChange*	Flag the beginning of the state for checking

(continued)

S. Vongbunyong and W.H. Chen, *Disassembly Automation*, Sustainable Production, Life Cycle Engineering and Management, DOI 10.1007/978-3-319-15183-0

Table A.2 (continued)

Category	Primitive action	Fluent	Description
Location utility (2) also for line and point	*setProdCoordinate*	*rect(x,y,x,y,z)*	Set a VOI for product coordinate {P}
	offsetContourXY	*rect(x,y,x,y,z)*	(2) Offset contour
	offsetContourDepth	*rect(x,y,x,y,z)*	(2) Offset contour vertically
	rectRoiIs	*rect(x,y,x,y,z)*	Specify the ROI
	rectCutLocationIs	*rect(x,y,x,y,z)* or *box(x,y,x,y,z,z)*	(2) Specify the cutting location
KB	*addSequencePlan*	*sequencePlan*	Add plans to KB
	recallDSP	–	Recall plans from KB
	feedCustomPlan	–	Proceed to the next operation plan in the list
	feedDspComponent	–	Proceed to next component
Human assistance	*skipComponent*	–	Skip treat current component
	newCompLocation	*rect(x,y,x,y,z)*	Locate the current component
	deactivate	–	Stop human demonstrating
	All primitive cutting from (1)	Primitive geometry	Demonstrate cutting at specific location

Table A.3 Fluent as constant parameters

Fluent	Value (mm)
maxBackCoverDeepOffset	3
maxPcbCoverDeepOffset	12
maxPcbDeepOffset	12
maxCarrierDeepOffset	6
maxScrewDeepOffset	3
minIncrementDepth	1
incrementDepth	2
maxIncrementDepth	3
minZ	1
maxZ	80

Appendix B
Graphic User Interface

The user uses the graphic user interface (GUI) to interact with the system for process control and demonstrations in the learning process. In regard to the demonstration, the GUI is designed for intuitive and interaction that allows the user to precisely demonstrate the commands and primitive cutting operation (see Figs. B.1 and B.2). The issue of 2D and 3D perception of the user is taken into

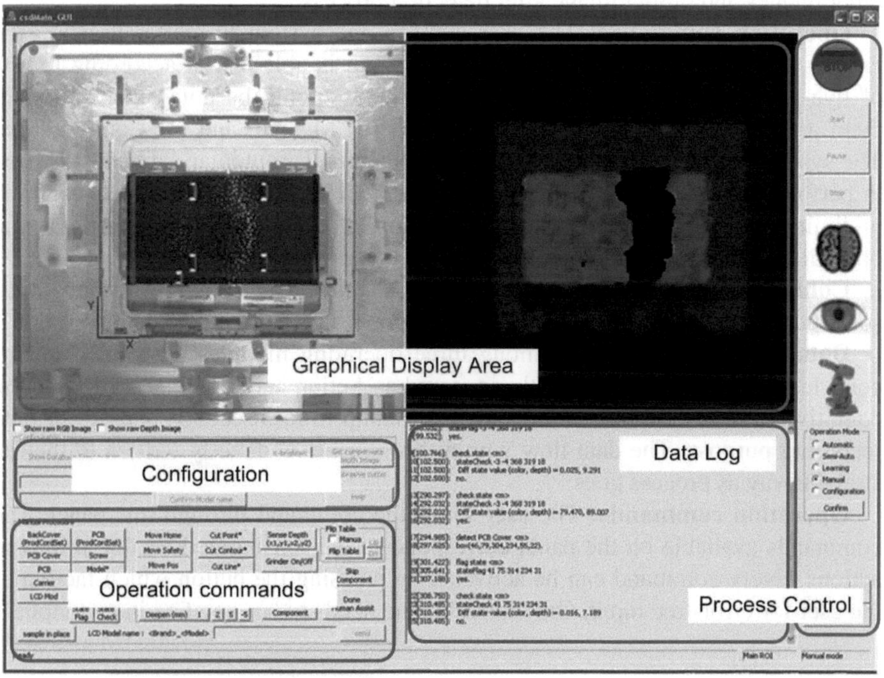

Fig. B.1 Graphical user interface console

S. Vongbunyong and W.H. Chen, *Disassembly Automation*, Sustainable Production, Life Cycle Engineering and Management, DOI 10.1007/978-3-319-15183-0

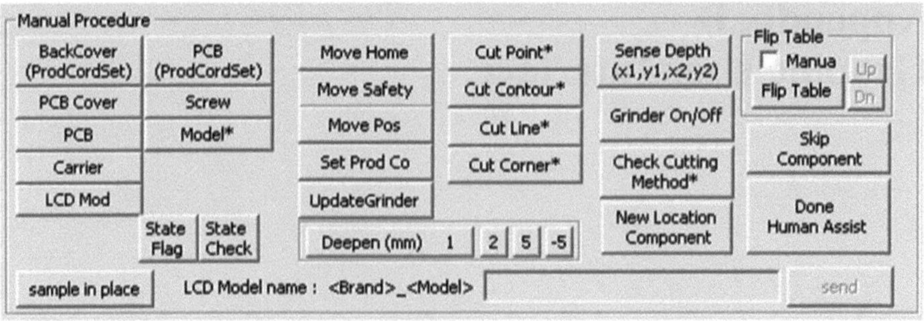

Fig. B.2 Graphical user interface—operation part

account. The GUI is developed and operated in C/C++ under Visual Studio 2008. The GUI consists of 5 main areas: (a) Graphic display area, (b) Operation commands, (c) Configuration, (d) Data log, and (e) Process control.

Graphic display area: Snapshots of a colour and a depth images snap captured during the process are rendered on this area. The image can be switched between input images and output image after detection process.

Operation commands: The user controls the system to start/pause/stop the process using this panel. In addition, the system is able to run according to one of five operation modes specified by the user. It should be noted that only 3 of them are available, including (a) Automatic, (b) Manual, and (c) Configuration. The system performs disassembly autonomously in the automatic mode. It is used in performance test in Chap. 6. The manual mode is used to test the concept and preliminary test, e.g. vision system's detection, of each function before the actual operation. The configuration mode is explained next.

Configurations: In the configuration mode, the user allows to minor adjust some parameters in regard to calibration purpose, e.g. depth image compensation.

Data log: The data flows among three operating modules are shown in this console in the form of text, mostly appeared as Action and Fluent according to the cognitive robotic module's command. Timestamp in milliseconds is used for data recording purpose. The data flow within this console is directly written to the file straightaway as process goes.

Operation commands: The user sends the command through this panel. The commands available on the panel correspond to the sensing actions and primitive actions. Every command can be activated by pressing the button which facilitates the user for error-free input. Only model name needs to be specified in text input.

Index

S. Vongbunyong and W.H. Chen, *Disassembly Automation*, Sustainable Production, Life Cycle Engineering and Management, DOI 10.1007/978-3-319-15183-0

MIX
Papier aus verantwortungsvollen Quellen
Paper from responsible sources
FSC® C105338

If you have any concerns about our products,
you can contact us on
ProductSafety@springernature.com

In case Publisher is established outside the EU,
the EU authorized representative is:
Springer Nature Customer Service Center GmbH
Europaplatz 3, 69115 Heidelberg, Germany

Printed by Libri Plureos GmbH
in Hamburg, Germany